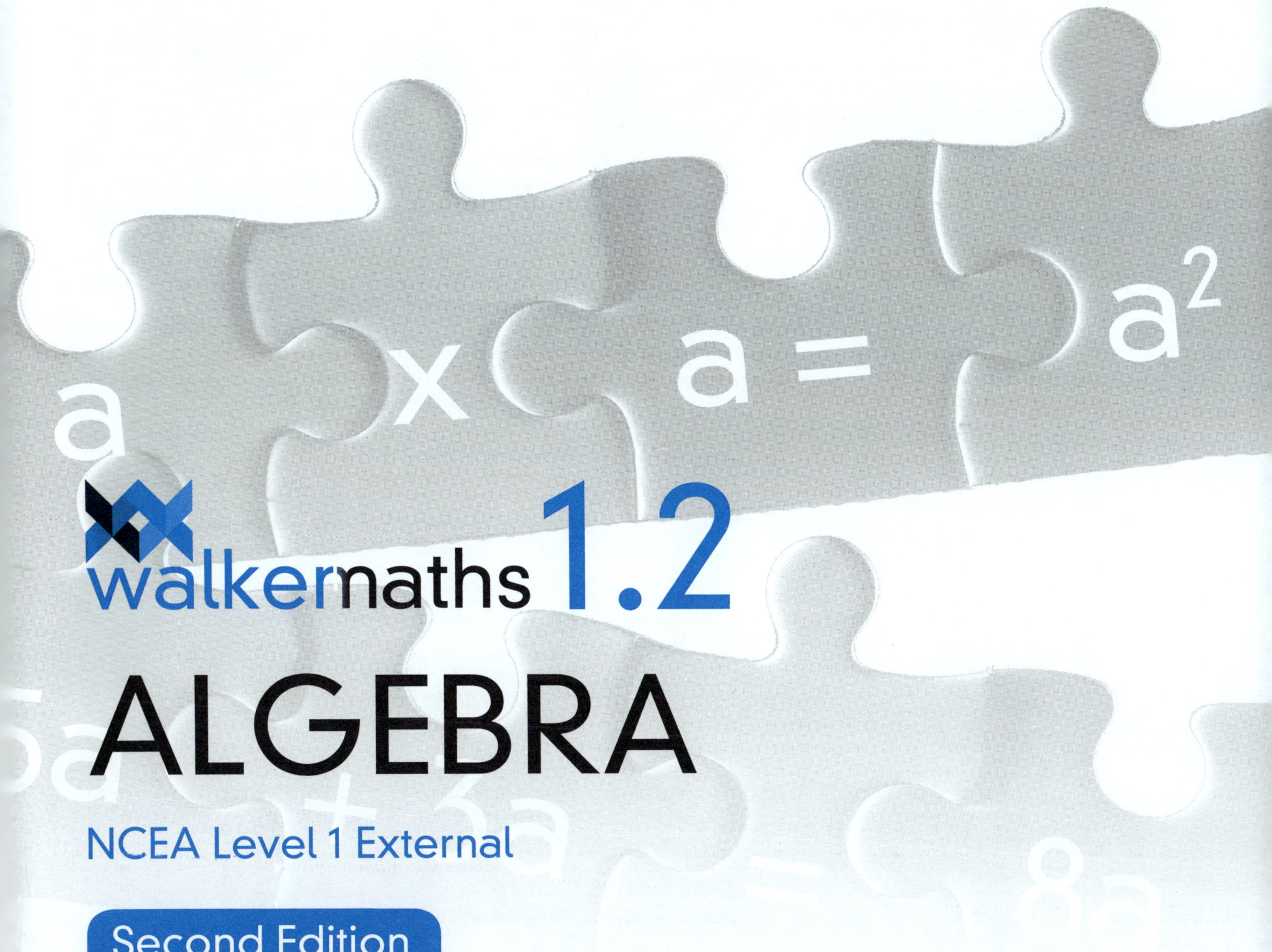

a × a = a²

walkermaths 1.2

ALGEBRA

5a + 3a = 8a

NCEA Level 1 External

Second Edition

Charlotte Walker and Victoria Walker

Walker Maths 1.2 Algebra
2nd Edition
Charlotte Walker
Victoria Walker

Editor: Eva Chan
Designer: Cheryl Smith, Macarn Design
Production controller: Alice Kane

Any URLs contained in this publication were checked for currency during the production process. Note, however, that the publisher cannot vouch for the ongoing currency of URLs.

Acknowledgements
Cover photo courtesy of Shutterstock.

For product information and technology assistance,
in Australia call **1300 790 853**;
in New Zealand call **0800 449 725**

For permission to use material from this text or product, please email **aust.permissions@cengage.com**

National Library of New Zealand Cataloguing-in-Publication Data
A catalogue record for this book is available from the National Library of New Zealand.

ISBN 978 0 17 043856 8

Cengage Learning Australia
Level 7, 80 Dorcas Street
South Melbourne, Victoria Australia 3205

Cengage Learning New Zealand
Unit 4B Rosedale Office Park
331 Rosedale Road, Albany, North Shore 0632, NZ

For learning solutions, visit **cengage.co.nz**

Printed in China by 1010 Printing International Limited
10 11 12 26

CONTENTS

ISBN: 9780170438568

 ISBN: 9780170438568

Glossary

Make your own glossary of key terms:

Term	Definition	Picture/Example
Expression		
Term		
Constant		
Power		
Index (plural: indices)		
Exponent		
Variable		
Coefficient		
Expand		
Factorise		

ISBN: 9780170438568

Term	Definition	Picture/Example
Solve		
Simplify		
Evaluate		
Denominator		
Numerator		
Quadratic		
Dimensions		
Product		
Sum		
Reciprocal		
Lowest Common Multiple (LCM)		

ISBN: 9780170438568

Simplifying expressions

Multiplying and dividing

The order within simplified expressions should be:

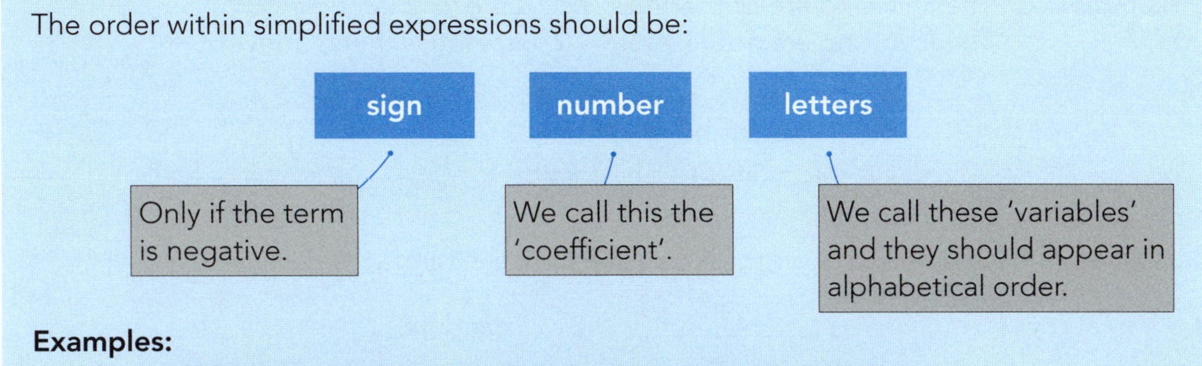

Examples:

Unsimplified expressions	Simplified expressions
$a \times 2$	$2a$
$a \times a \times a$	a^3
$3a \times 2a$	$6a^2$
$-5 \times -a \times b$	$5ab$
$a \times d \times 2c \times -4b$	$-8abcd$
$a \div 4$	$\frac{a}{4}$ or $\frac{1}{4}a$

Index form or power form.

Simplify the following expressions.

1 $y \times y \times y =$ ____________ 2 $d \times f \times a =$ ____________

3 $7 \times d =$ ____________ 4 $2d \times 3f =$ ____________

5 $-e \times 3f \times g =$ ____________ 6 $d \times f \times f \times 7 =$ ____________

7 $5a \times -2a =$ ____________ 8 $-4b \times 2a \times 3c =$ ____________

9 $5dc \times -d =$ ____________ 10 $-1 \times 3f \times -6 =$ ____________

11 $g \div 2 =$ ____________ 12 $14d \div e =$ ____________

13 $3 \div e =$ ____________ 14 $20dg \div f =$ ____________

15 $(3d \times 4f) \div ac =$ ____________ 16 $12efg \div (a \times a \times a) =$ ____________

ISBN: 9780170438568

Adding and subtracting

- Terms can only be added or subtracted if they are 'like' terms.
- 'Like' terms must have exactly the same variables, and each variable must be raised to exactly the same power.
- Order does not matter.
- The sign does not matter.

Examples: The following **are** like terms: a^2b^5c, $2a^2b^5c$, $b^5a^2c^1$, $-a^2cb^5$
The following are **not** like terms (compared with the above):

a^2b^6c — the 'b' is raised to the power of 6, not 5
b^5c — there is no 'a' term
a^2b^5cd — includes a 'd' term

State whether each of the following pairs are like or unlike terms.

1 xy and yx ____________

2 $5x$ and $3x^2$ ____________

3 ab^2c and c^2ab ____________

4 x and $-2x$ ____________

5 xy^2 and $-0.1y^2x$ ____________

6 7 and 1 ____________

7 x^1 and x ____________

8 $4cd$ and $-9dce$ ____________

9 x^2yz and $-y^2zx$ ____________

10 $w^5x^3y^2z^1$ and $-y^2zw^5x^3$ ____________

- When adding or subtracting, you can combine only like terms.

Examples:
1 $7a + 2a - a^2 = 9a - a^2$
2 $10 + 5y - 3 - y = 7 + 4y$
3 $b^4 + 6b^4 - b^3 = 7b^4 - b^3$
4 $2ab^2c - a^2bc - 2bca^2 = 2ab^2c - 3a^2bc$ or $-3a^2bc + 2ab^2c$

The **order** of the terms in your answer does not matter.

Simplify these by adding or subtracting like terms.

11 $x + x + x + x =$ ____________

12 $5x^2 + x^2 + x =$ ____________

13 $12 + 3ab - 7 - ab =$ ____________

14 $7p^5q - pq^5 + 2p^5q + 4pq^5 =$ ____________

15 $2y^3 + 2y^2 + 6y^2 - 3y^3 =$ ____________

16 $x^2 + 5x + 7x + 35 =$ ____________

17 $11x + 6 - 4y - 5 =$ ____________

18 $6ab + 4a - 3b =$ ____________

19 $-5p + 17 - p + q =$ ____________

20 $x^2 - 9x - 3x + 27 =$ ____________

21 $y^2 - y + 8y - 8 =$ ____________

22 $3x^2 + 7xy - 8y^2 - xy + y^2 =$ ____________

23 $x^3 + x^2 - x^3 =$ ____________

24 $6x^2yz + 2xyz^2 - x^2yz - xyz^2 =$ ____________

ISBN: 9780170438568

Exam questions

Answer the following questions. In all cases, simplify your answers where possible.

1 The lengths of the sides of a triangle are $3y$, $y + 2$ and $y^2 - 5$. Write an expression for the perimeter of the triangle.

____

____

2 Another triangle has a perimeter of $3x^2 + z + 1$. If two of the sides have lengths x^2 and $4z$, write an expression for the third side.

____

____

3 The figure drawn below is made up of two rectangles. Write expressions for the lengths of each of the blue sides, and the perimeter.

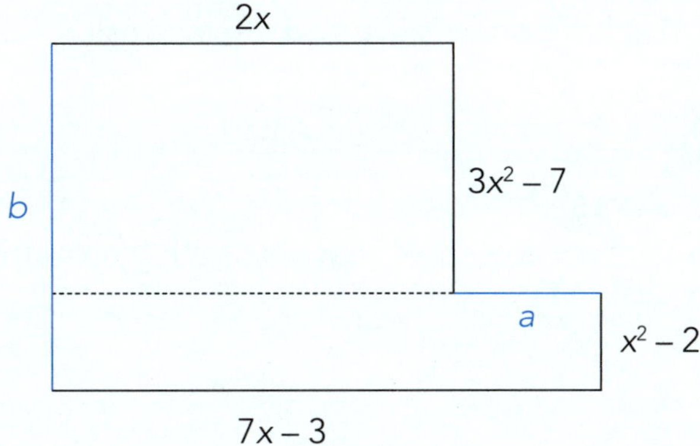

a = ______________________________

= ______________________________

b = ______________________________

= ______________________________

Perimeter = ______________________________

= ______________________________

4 A pentangle has sides with lengths xy, x^2, $7y^2$ and $6x$.
If its perimeter is $2x^2 + 5xy + y^2 + x$, write an expression for its fifth side.

____

____

ISBN: 9780170438568

Powers

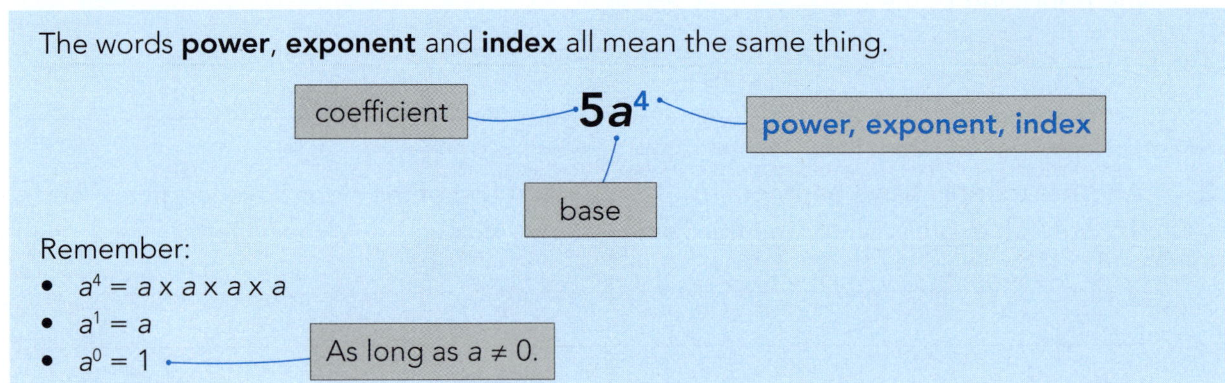

Multiplying terms that have powers

When multiplying terms that have powers, we **add** the indices.

Examples:

1 $a^5 \times a^2 = a^7$ — $5 + 2 = 7$

2 $3a^4 \times 5a^9 = 15a^{13}$ — Always deal with the **coefficients** first.

3 $2ab \times 7ab^2 = 14a^2b^3$

Simplify these.

1 $p \times p \times p \times p \times p$

2 $a^4 \times a$

3 $b^7 \times b^2$

4 $g^3 \times g \times g^5$

5 $3a^2 \times 2a$

6 $2ab \times 5a^2b$

7 $15b^{10} \times 2b^5$

8 $fg^2 \times f \times g^4 \times fg$

9 $(-5y^6) \times (-3y^4)$

10 $(-a^2) \times (-b^3) \times a \times (-b)$

11 $-2y^4 \times 3y^2$

12 $2z^2 \times 4z^2 \times z^6$

ISBN: 9780170438568

Dividing terms that have powers

When dividing terms that have powers, we **subtract** the indices.

Examples:

1 $a^5 \div a^2 = a^3$ or $\frac{a^6}{a^2} = a^4$ — $6 - 2 = 4$

2 **12**$a^6 \div$ **6**$a^2 =$ **2**a^4 — Always deal with the **coefficients** first.

3 $\frac{14x^2}{7x^5} = \frac{2}{x^3}$

Simplify these.

1 $\frac{y^8}{y^2}$

2 $\frac{y^4}{y}$

3 $b^{12} \div b^3$

4 $\frac{10b^{10}}{2b^5}$

5 $\frac{15y^4}{30y}$

6 $\frac{x^2y^4}{y^3}$

7 $\frac{8y^4z}{2y^2}$

8 $\frac{-x^4}{2x^5}$

9 $\frac{12xy^2}{4y^7}$

10 $\frac{-2x^4}{6x^3}$

11 $\frac{-x^4y}{-3x^5y}$

12 $\frac{-10x^4y^2z^3}{-5x^7z}$

13 $3xy \div 18xy$

14 $\frac{7xyz^2}{14x^2yz^3}$

ISBN: 9780170438568

Powers of powers

When finding a power of a power, we **multiply** the indices.

Examples:

1 $(a^2)^3 = a^2 \times a^2 \times a^2$
$= a^6$ — $2 \times 3 = 6$

2 $(4a^3)^2 = 4^2 \times (a^3)^2$
$= 16a^6$

or $(4a^3)^2 = 4a^3 \times 4a^3$
$= 16a^6$

3 $\left(\frac{1}{2}a^4\right)^3 = \left(\frac{1}{2}\right)^3 \times (a^4)^3$
$= \frac{1}{8}a^{12}$ or $\frac{a^{12}}{8}$ or $0.125a^{12}$

or $\left(\frac{1}{2}a^4\right)^3 = \left(\frac{1}{2}a^4\right) \times \left(\frac{1}{2}a^4\right) \times \left(\frac{1}{2}a^4\right)$
$= \frac{1}{8}a^{12}$ or $\frac{a^{12}}{8}$ or $0.125a^{12}$

Simplify these.

1 $(b^4)^2$

2 $(2x^3)^4$

3 $(-a^7)^2$

4 $(-y^3)^5$

5 $(-3x^4)^2$

6 $(-4x^4)^3$

7 $-2(5x^3)^2$

8 $(x^2 y^4 z^3)^3$

9 $-(-3a^3 b^2)^2$

10 $2(-3x y^3)^2$

11 $\left(\frac{1}{2}x^2\right)^3$

12 $\left(-\frac{1}{5}x^4\right)^2$

ISBN: 9780170438568

Mixing it up

Examples:

1 $\dfrac{a^2 \times a^4}{a \times a^2} = \dfrac{a^6}{a^3}$
$= a^3$

2 $\dfrac{(2a^2)^3}{2a^2} = \dfrac{8a^6}{2a^2}$
$= 4a^4$

3 $\dfrac{(a^2)^3}{(3a^4)^2} = \dfrac{a^6}{9a^8}$
$= \dfrac{1}{9a^2}$

4 $(y^3)^2\,(2x)^3 = y^6 \times 8 \times y^3$
$= 8y^9$

Simplify these.

1 $b^6 \times b^2 \div b^4$

2 $\dfrac{x^2\,y^2}{(xy)^2}$

3 $\dfrac{c^2b \times cb^3}{cb}$

4 $\dfrac{(3b)^2}{b^3}$

5 $\left(\dfrac{a^2}{2}\right)^3$

6 $\left(\dfrac{4z^3}{5z}\right)^2$

7 $\left(\dfrac{4z^5}{z}\right)^3$

8 $\left(\dfrac{zy}{5zy^4}\right)^2$

9 $(3y^2z^4)^3 \times 2yz$

10 $\dfrac{(4z^5)^3y}{y^2z}$

ISBN: 9780170438568

Roots

- Finding a root is the reverse of finding a power.
- Roots are written as fractional powers.

Roots of numbers:

- $\sqrt{9}$ is the same as writing $\sqrt[2]{9}$ or $9^{\frac{1}{2}}$.
- $\sqrt[3]{64}$ is the same as writing $64^{\frac{1}{3}}$.

$\sqrt{9}$ is the same as $\sqrt{3 \times 3} = \sqrt{3^2} = 3$.

$\sqrt[3]{64}$ is the same as $\sqrt[3]{4 \times 4 \times 4} = \sqrt[3]{4^3} = 4$.

Roots of variables:

- $\sqrt{a^2}$ is the same as writing $\sqrt[2]{a^2}$ or $a^{\frac{2}{2}} = a$.
- $\sqrt[3]{a^3}$ is the same as writing $\sqrt[3]{a^3} = a^{\frac{3}{3}} = a$.

To find the **square** root of a power, you must **halve the power**: $\sqrt{a^{10}} = a^5$

$10 \div 2 = 5$

Examples:

1 $\sqrt{9x^8} = 3x^4$

2 $\sqrt{100x^6} = 10x^3$

Simplify these.

1 $\sqrt{25}$

2 $\sqrt{x^{10}}$

3 $\sqrt[3]{1000}$

4 $\sqrt{36x^6}$

5 $\sqrt{16x^{16}}$

6 $\sqrt[3]{64}$

7 $\sqrt{100x^{100}}$

8 $\sqrt{49x^8y^6}$

9 $\sqrt{(9x^{64})^2}$

10 $\sqrt{(2x^2y^4)^6}$

11 $\sqrt{25x^{10}y^6}$

12 $\sqrt{64x^{64}y^{16}z^{36}}$

 ISBN: 9780170438568

Exam questions

Answer the following questions. In all cases, simplify your answers where possible.

1 One side of a rectangle is $4a^2bc$ units long, and the other side is $3ab^3$ units long. Write an expression for the area of the rectangle.

Area = ______________________________

____

2 A square has sides which are $3p^4q$ units long. Write an expression for the area of the square.

Area = ______________________________

____

3 The cuboid shown in the diagram has sides with the lengths shown. Write an expression for the volume of the cuboid.

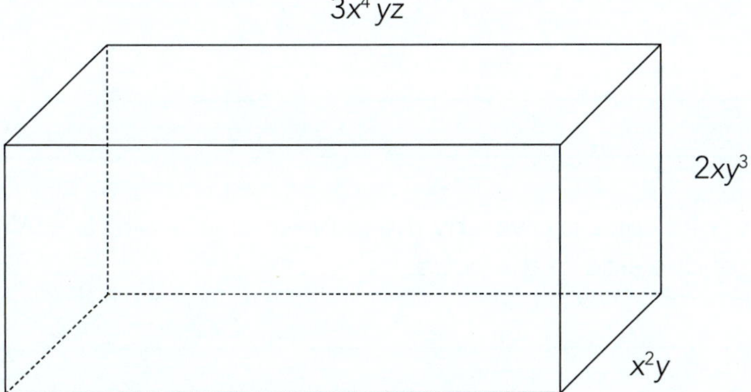

Volume = ______________________________

____

4 The area of a square is given by the expression Area = $16x^{16}y^4$. Write an expression for the length of one side.

Length = ______________________________

____

5 The area of a rectangle is given by the expression Area = $36x^8y^4$. If one of its sides has the length $9x^2y$, write an expression for the length of the other side.

Length = ______________________________

____

ISBN: 9780170438568

6 A cube has sides which are $5x^2y$ long. Write an expression for the volume of the cube.

Volume = ______________________________

____

7 The lengths of two sides of a cuboid are shown on the diagram. If the volume of the cuboid is $36x^8y^{12}z$, write an expression for the length (L) of the third side.

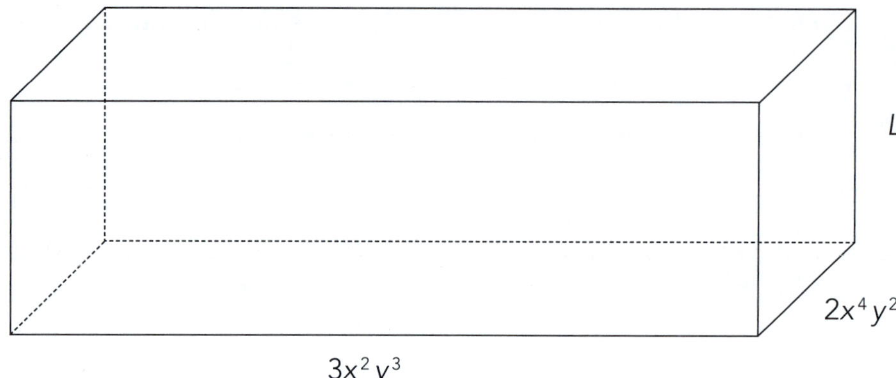

Length = ______________________________

____

8 The volume of a cube is given by the expression $V = 64a^6b^{12}$. Write an expression for the length of one side of the cube.

Length = ______________________________

____

9 The area of one end of a cuboid is given by the expression Area = $16a^4b^6$. If the volume of the cuboid is given by the expression $V = 8a^{12}b^{24}$, write an expression for the length of the cuboid.

Length = ______________________________

____

10 The area of the end of a cylinder is given by the expression Area = πf^4g^2. Its length is fg^3. Write an expression for the volume of the cylinder. You may leave π in your answer.

Volume = ______________________________

____

ISBN: 9780170438568

Expanding and factorising

Expanding brackets

- In algebra, 'expand' means multiply out all the brackets.
- After expanding, you are expected to collect the like terms in order to simply the expression.

Examples: Expand and simplify the following.

1 $-2(5x + 3) = -10x - 6$

2 $4x(3 - 5x) = 12x - 20x^2$

3 $9 - 4(7x - 2) = 9 - 28x + 8$
$= -28x + 17$

Be careful!

4 $3(6x - 5) - (5x + 3) = 18x - 15 - 5x - 3$
$= 13x - 18$

Expand and simplify the following.

1 $5(3x - 4)$

2 $6x(2 - x)$

3 $-2x(3x - 5)$

4 $7(3x - 4) + 12$

5 $1 - 2(4 - 3x) - 5$

6 $10x - 3(7 - 2x)$

7 $4(3 - 2x) + 2(x - 5)$

8 $2(9x + 4) + 3(5 + 2x)$

9 $x(2x - 7) - (x - 4)$

10 $9(2x - 1) - 5(3 - x)$

11 $5x^2 - 2x(3 - 4x) + x^2$

12 $x(1 - 2x) - 7x(3x + 4)$

ISBN: 9780170438568

Factorising

- Factors are terms that are **multiplied** together (rather than added or subtracted). For example, 2 and 3 are factors of 6 because 2 x 3 = 6.
- Expressions with brackets are usually in factorised form.

Examples:

In each factorised expression there is an unwritten **x (times)** sign before the bracket.

Factorised form	Unfactorised (expanded) form
$2(3x + 4)$	$6x + 8$
$x(5 - 3x)$	$5x - 3x^2$
$3(2x^2 + x - 7)$	$6x^2 + 3x - 21$

- When factorising, you must factorise **completely**. There must be no common factor for the terms inside the brackets.
- You need to ask yourself: '**What is the biggest thing (or things) that will divide into every term?**'
- Do not use fractions when factorising.

The word '**completely**' is understood, but not written here.

Examples: Factorise .. the following.

1 $12x - 18 = 6(2x - 3)$

Notice that $2(6x - 9)$ is not a correct answer because 3 can divide into both 6x and 9.
Similarly, $3(4x - 6)$ is not a correct answer because 2 divides into both 4x and 6.

2 $20x^2 + 30x = 10x(2x + 3)$

All of the following are true but **incorrect** answers:

$2(10x^2 + 15x)$	$5(4x^2 + 6x)$	$10(2x^2 + 3x)$
$2x(10x + 15)$	$5x(4x + 6)$	$\frac{1}{2}(40x^2 + 60x)$

Circle the correct factorised answer for each of the following.

1 $16x + 8$

$2(8x + 4)$
$1(16x + 8)$
$8(2x + 1)$
$4(4x + 2)$
$\frac{1}{2}(32x + 16)$

2 $36x - 60$

$6(6x - 10)$
$1(36x - 60)$
$12(3x - 5)$
$3(12x - 20)$
$24(1.5x - 2.5)$

ISBN: 9780170438568

3	$100 + 20x$	$2(50 + 10x)$ $\frac{1}{2}(200 + 40x)$ $100\left(1 + \frac{1}{5}x\right)$ $20(5 + x)$ $10(10 + 2x)$	**4**	$60x - 24x^2$	$x(60 - 24x)$ $6(10x - 4x^2)$ $6x(10 - 4x)$ $12(5x - 2x^2)$ $5(3x - 4x^2)$ $12x(5 - 2x)$
5	$12xy^2 + 4xy$	$4xy(3y + 1)$ $4x(3y^2 + y)$ $2xy(6y + 2)$ $4y(3xy + x)$ $xy(12y + 4)$	**6**	$ab^4c^3 + a^2b^3c^4$	$ab^2c^3(b^2 + ab)$ $b^3c^3(ab + a^2c)$ $ab^3c^2(bc + ac^2)$ $ab^3c^3(b + ac)$ $ab^2c^2(b^2c + abc^2)$

Factorise the following.

7 $5x - 15$

____

8 $x^2 - 3x$

____

9 $8x + 16$

____

10 $24 - 18a$

____

11 $10y + 6$

____

12 $x^2 + 11x$

____

13 $2x + 3x^2$

____

14 $5x^2 + 15x$

____

15 $4y^2 + 12y$

____

16 $4a^2 + 2a$

____

17 $3x^2 + 15x$

____

18 $21y - 14y^3$

____

19 $10a^3 + 2a^2$

____

20 $12b^2 + 16b^3$

____

21 $xy + 9x^2y^2$

____

22 $14x^2y + 7xy^2$

____

23 $2ab^2c - a^2bc^3$

____

24 $24ab^2c^3 - 8a^2bc^3$

____

25 $10xy + 5x^2 - xy^2$

____

26 $12xy^2 + 4x^2y - 8x^2y^2$

____

ISBN: 9780170438568

Exam questions

Answer the following questions. In all cases, simplify your answers where possible.

1 A rectangle has sides with the lengths $11p$ and $2(5p - 3)$. Write an expression for the perimeter of the rectangle.

Perimeter = ____________________

____

2 A triangle has sides with lengths $9t$, $4(2t - 3)$ and $5(t + 6)$ Write an expression for the perimeter of the triangle.

Perimeter = ____________________

____

3 A triangle has two sides with lengths $3(x - 1)$ and $7x$. It its perimeter is $14x - 1$, write an expression for the length of the third side.

Length = ____________________

____

4 A rectangle has one side with a length of $2a^3$. It its area is $6a^3 - 10a^4$, write an expression for the length of the other side.

Length = ____________________

____

5 A rectangle has one side with a length $7(1 - 2y)$. If its area is given by the expression $21x^2y - 42x^2y^2$, write an expression for the length of the other side.

Length = ____________________

____

6 Find the mean of these three terms: $2(4x^2 - 3)$; $5x - 4$; $x^2 + x - 2$.

Mean = ____________________

____

7 Find the mean of these four terms: $9xy - y^2$; $2 - yx$; $5y^2 - 7 - 2y$; $5 - 10y$.

Mean = ____________________

____

ISBN: 9780170438568

Algebraic fractions

Multiplying fractions

$$\frac{a}{b} \times \frac{c}{d} = \frac{ac}{bd}$$

Multiply the numerators and multiply the denominators.

Examples:

1 $\frac{a}{3} \times \frac{2}{5} = \frac{2a}{15}$

2 $\frac{2z}{7} \times \frac{z}{3} = \frac{2z^2}{21}$

3 $\frac{y}{2z} \times \frac{z}{5y} = \frac{1}{10}$

4 $\frac{2(y+1)}{4} \times \frac{3}{5} = \frac{6(y+1)}{20} = \frac{3(y+1)}{10}$

Try for yourself.

1 $\frac{y}{2} \times \frac{4}{9} =$

2 $\frac{7}{z} \times \frac{2}{z} =$

3 $\frac{2z}{7} \times \frac{4}{5} =$

4 $\frac{3x}{y} \times \frac{6}{y} =$

5 $\frac{5y}{6z} \times \frac{1}{y} =$

6 $\frac{1}{4} \times \frac{2y}{3} =$

7 $\frac{3}{4} \times \frac{y-5}{5} =$

8 $\frac{2yz}{7} \times \frac{yz^2}{2} =$

9 $\frac{2z}{3} \times \frac{4}{5z} \times \frac{z}{2} =$

10 $\frac{3(x+1)}{2} \times \frac{x}{5} =$

11 $\frac{6x^2yz}{5} \times \frac{yz}{3x} =$

12 $\frac{3z}{y-1} \times \frac{y}{6z} =$

ISBN: 9780170438568

Dividing fractions

$$\frac{a}{b} \div \frac{c}{d} = \frac{a}{b} \times \frac{d}{c} = \frac{ad}{bc}$$

÷ → x

Change the ÷ sign to a x sign and flip the second fraction.

invert the second fraction → the reciprocal

Examples:

1 $\frac{z}{8} \div \frac{z}{2} = \frac{z}{8} \times \frac{2}{z}$

$= \frac{1}{4}$

2 $\frac{6y}{5} \div \frac{3}{y} = \frac{6y}{5} \times \frac{y}{3}$

$= \frac{2y^2}{5}$

Simplify these.

1 $\frac{12}{x} \div \frac{2}{x} =$

2 $\frac{1}{x} \div \frac{1}{y} =$

3 $\frac{3}{x} \div \frac{2}{x} =$

4 $\frac{9y}{x} \div \frac{3}{x} =$

5 $\frac{5x}{2} \div \frac{x}{10} =$

6 $\frac{4x}{5} \div \frac{1}{15x} =$

7 $\frac{x}{y^2} \div x =$

8 $\frac{15y}{2x} \div \frac{3y}{x} =$

9 $\frac{3x^2}{y} \div \frac{5x}{4y} =$

10 $\frac{10x^2}{y^2} \div \frac{2x}{5y} =$

 ISBN: 9780170438568

Adding and subtracting fractions

1 Where the denominators are the same

Adding or subtracting fractions with the **same denominator**:

$$\frac{a}{b}+\frac{c}{b}=\frac{a+c}{b} \text{ or } \frac{a}{b}-\frac{c}{b}=\frac{a-c}{b}$$

Examples:

1 $\frac{5x}{3}+\frac{2x}{3}=\frac{7x}{3}$

2 $\frac{4}{x}+\frac{3}{x}=\frac{7}{x}$

3 $\frac{x}{8}+\frac{3x}{8}=\frac{4x}{8}$

$=\frac{x}{2}$

4 $\frac{6x}{7}-\frac{x+2}{7}=\frac{6x-(x+2)}{7}$

$=\frac{5x-2}{7}$

Try for yourself.

1 $\frac{x}{9}+\frac{4x}{9}=$

2 $\frac{7}{x}+\frac{2}{x}=$

3 $\frac{5x}{7}-\frac{4x}{7}=$

4 $\frac{7x}{y}-\frac{x}{y}=$

5 $\frac{2x}{11}+\frac{4x}{11}-\frac{x}{11}=$

6 $\frac{3x}{10}+\frac{2x}{10}=$

7 $\frac{x}{5}+\frac{4x}{5}=$

8 $\frac{7}{x}+\frac{4}{x}-\frac{5}{x}=$

9 $\frac{2x-1}{7}+\frac{4x}{7}=$

10 $\frac{5x+2}{9}+\frac{3(2x-1)}{9}=$

11 $\frac{5(2x-3)}{12}-\frac{4x-3}{12}=$

12 $\frac{6(3-2x)}{13}-\frac{x+5}{13}=$

ISBN: 9780170438568

2 Where the denominators are different

Where fractions have **different denominators**, multiply both the numerator and the denominator by a number (or variable) that will create equal denominators.

$$\frac{a}{b} + \frac{c}{d} = \frac{a}{b} \times \frac{d}{d} + \frac{c}{d} \times \frac{b}{b}$$
$$= \frac{ad}{bd} + \frac{cb}{db}$$
$$= \frac{ad + bc}{bd}$$

$\frac{d}{d} = 1$ and $\frac{b}{b} = 1$

Examples:

1 $\frac{x}{8} + \frac{x}{3} = \frac{x}{8} \times \frac{3}{3} + \frac{x}{3} \times \frac{8}{8} = \frac{3x}{24} + \frac{8x}{24}$
$= \frac{11x}{24}$

2 $\frac{x}{2} + x = \frac{x}{2} + \frac{x}{1} \times \frac{2}{2}$
$= \frac{x}{2} + \frac{2x}{2} = \frac{3x}{2}$

3 $\frac{2x}{5} - \frac{1}{4} = \frac{2x}{5} \times \frac{4}{4} - \frac{1}{4} \times \frac{5}{5} = \frac{8x}{20} - \frac{5}{20}$
$= \frac{8x - 5}{20}$

4 $\frac{2x + 3}{2} - \frac{x}{3} = \frac{2x + 3}{2} \times \frac{3}{3} - \frac{x}{3} \times \frac{2}{2}$
$= \frac{6x + 9 - 2x}{6} = \frac{4x + 9}{6}$

Simplify these.

1 $\frac{2x}{3} + \frac{x}{4}$

2 $\frac{3x}{7} + \frac{x}{2}$

3 $\frac{2x}{3} - \frac{x}{5}$

4 $\frac{x}{2} - \frac{3x}{7}$

5 $\frac{xy}{3} - \frac{1}{5}$

6 $4 - \frac{x}{6}$

7 $\frac{y}{4} - \frac{x}{6}$

8 $3 + \frac{x - 2}{5}$

9 $\frac{2x - 5}{5} + x$

10 $\frac{2x - 1}{3} - \frac{1}{2}$

ISBN: 9780170438568

3 Where the denominators are different algebraic expressions

Where fractions have different denominators which are **algebraic expressions**, use the same process.

$$\frac{a}{b}+\frac{c}{d}=\frac{a}{b}\times\frac{d}{d}+\frac{c}{d}\times\frac{b}{b}$$
$$=\frac{ad}{bd}+\frac{cb}{db}$$
$$=\frac{ad+bc}{bd}$$

$\frac{d}{d}=1$ and $\frac{b}{b}=1$

Examples:

1 $\frac{2}{x-3}+\frac{1}{x+4}=\frac{2}{x-3}\times\frac{x+4}{x+4}+\frac{1}{x+4}\times\frac{x-3}{x-3}$
$=\frac{2(x+4)+1(x-3)}{(x-3)(x+4)}$
$=\frac{2x+8+x-3}{(x-3)(x+4)}$
$=\frac{3x+5}{(x-3)(x+4)}$

2 $\frac{x+1}{2x+5}-\frac{x-4}{3x-1}=\frac{x+1}{2x+5}\times\frac{3x-1}{3x-1}+\frac{x-4}{3x-1}\times\frac{2x+5}{2x+5}$
$=\frac{(x+1)(3x-1)-(x-4)(2x+5)}{(2x+5)(3x-1)}$
$=\frac{(3x^2-x+3x-1)-(2x^2+5x-8x-20)}{(3x-1)(2x+5)}$
$=\frac{3x^2+2x-1-2x^2+3x+20}{(3x-1)(2x+5)}$
$=\frac{x^2+5x+19}{(3x-1)(2x+5)}$

Simplify these.

1 $\frac{2}{x+1}+\frac{3}{2x-4}$

2 $\frac{3}{3x-2}-\frac{x}{x+1}$

3 $\frac{x+3}{x+1}+\frac{x-1}{x+5}$

4 $\frac{x+1}{3x-2}+\frac{2x-3}{x+4}$

ISBN: 9780170438568

Exam questions

Answer the following questions. In all cases, simplify your answers where possible.

1 If $\frac{4x^2y}{3z} \times \frac{a}{b} = \frac{4x}{z^2}$, write expressions for a and b.

2 Chris said that $\frac{8p}{5q} \div \frac{p^2}{10} = \frac{16p}{q}$. Is he correct? Explain your reasoning.

3 If $\frac{15}{q} \div \frac{a}{b} = 6pq^2$, write expressions for a and b.

4 If $\frac{4(y + 3)}{x} - \frac{a}{b} = \frac{y + 17}{x}$, write expressions for a and b.

5 A shed is built on the side of a garage. The area of the garage is $\frac{7x^2}{3}$ m². The area of the shed is $\frac{2x}{5}$ m². What is the total area written as a single fraction?

6 Aroha said that $\frac{2}{x} + \frac{3(1 - x)}{y} = \frac{2y + 3x - 3x^2}{xy}$. Is she correct?

7 If $\frac{5}{x - 1} + \frac{x + 1}{2x - 3} = \frac{a}{(x - 1)(2x + 3)}$, write an expression for a.

 ISBN: 9780170438568

Substitution

- When substituting, you replace variables with numbers.
- It is important to remember BEDMAS when doing this.

Examples:

1 If $E = \frac{1}{2}mv^2$, $m = 20$ and $v = 12$, calculate the value of E.

$E = \frac{1}{2} \times 20 \times 12^2$,

$= 1440$

Only the 12 is squared.

2 If $A = \frac{\sqrt{b+c}}{d}$, $b = 81$, $c = 19$ and $d = 2$, calculate the value of A.

$A = \frac{\sqrt{81 + 19}}{2}$

$= 5$

Don't forget to do the 81 + 19 first.

If $b = 6$, $c = 3$, $d = -4$ and $e = -2$, calculate the values of A.

1 $A = (b - c)(c + d)$

2 $A = \frac{b}{2}(b - c)$

3 $A = \frac{1}{2}bd^2$

4 $A = \frac{b - d}{e}$

5 $A = \frac{\sqrt{bc + e}}{e}$

6 $A = b^2 - 5b - 7$

7 $A = d^2 - 7d - 13$

8 $A = \sqrt{b^2 - 4de}$

If $r = 3$ and $h = 4$, calculate the values of V. You may leave π in your answer.

9 $V = h^3 - r^3$

10 $V = \pi r^2 h$

11 $V = \frac{4}{3}\pi r^3$

12 $V = r^3 - \frac{1}{3}\pi r^3$

ISBN: 9780170438568

Solving linear equations

'Solve' means 'find a value for x'.

Rules:
1. You can do anything you like to an equation as long as you do the **same to both sides**.
2. There should be only **one equals sign** per line.
3. Collect all the variables on one side and numbers on the other side.
4. When you want to get rid of something, perform the **opposite** operation.
5. Your answer should always be in the form **x = ...**

Trick: If you need to change the sign of everything, multiply **both** sides by **-1**.

One-step equations

Examples:

1. $y + 17.2 = 11$ **(– 17.2)**
   $y = -6.2$
2. $1.23 - x = 4$ **(– 1.23)**
   $-x = 2.77$ **(x by -1)**
   $x = -2.77$
3. $-6x = 20$ **(÷ by -6)**
   $x = -3.\dot{3}$
4. $\frac{x}{5} = -1.4$ **(x by 5)**
   $x = -7$

Solve the following.

1 $x - 0.6 = 1.1$

2 $x + 6.9 = 10$

3 $5x = 14$

4 $\frac{x}{8} = 0.25$

5 $x - 56 = -103$

6 $0.52 + x = -0.16$

7 $17.2 - x = 5.7$

8 $-1\frac{1}{2}x = -0.9$

9 $-x + 0.3 = 0.2$

10 $-143 - x = -10$

ISBN: 9780170438568

Two-step equations

Tricks: **1** Do the adding or subtracting **before** the multiplying or dividing.
2 If the variable (e.g. x) is on the right, **swap** the sides using the '=' sign as a centre.

e.g. $10.5 = 8x + 0.9 \longrightarrow 8x + 0.9 = 10.5$

Examples:

1
$10.5 = 8x + 0.9$
$8x + 0.9 = 10.5$ **(– 0.9)**
$8x = 9.6$ **(÷ by 8)**
$x = 1.2$

2
$5 - 2.5x = -12.5$ **(– 5)**
$-2.5x = -17.5$ **(÷ by -2.5)**
$x = 7$

Solve the following.

1 $5x - 3 = 16$

2 $0.5x + 9 = 14$

3 $23 + 1.2x = 26.6$

4 $168 - 4x = 1$

5 $0.06x - 3.6 = -4.8$

6 $110x + 7.9 = 3.5$

7 $25x - 5 = 0$

8 $213 = 7x + 101$

9 $-5.4 = 0.2x - 14.8$

10 $-0.009 = 0.006 - 3x$

ISBN: 9780170438568

Equations with a variable on both sides

Remember — collect all the terms that include an 'x' on one side and the constants on the other.

Examples:

1

$1 - 18x = 100 + 4x$ **(– 4x)**

$1 - 22x = 100$ **(– 1)**

$-22x = 99$ **(÷ by -22)**

$x = -4.5$

2

$9x + 1.1 = 2.7 - 7x$ **(+ 7x)**

$16x + 1.1 = 2.7$ **(– 1.1)**

$16x = 1.6$ **(÷ by 16)**

$x = 0.1$

Solve the following.

1 $5x - 3 = 4x + 7$

2 $13 - 5x = 10x + 43$

3 $2x - 19 = 5x - 46$

4 $1 - 5x = x + 31$

5 $9x - 150 = 1.5x$

6 $21x - 18 = 8x + 21$

7 $1 - 27x = -39 - 19x$

8 $3 - x = 5x + 735$

9 $8x + 17 = 11x + 5$

10 $12x + 0.2 = -0.3 + 7x$

ISBN: 9780170438568

Equations with brackets

Expand the brackets, then solve as you did in the last exercise.

Examples:

1 $5(2x - 1) = 2(4 + x)$
$10x - 5 = 8 + 2x$ **(– 2x)**
$8x - 5 = 8$ **(+ 5)**
$8x = 13$ **(÷ by 8)**
$x = 1.625$

2 $12(2 - x) = 3(5 - x) - 7(x + 1)$
$24 - 12x = 15 - 3x - 7x - 7$
$24 - 12x = 8 - 10x$ **(+ 10x)**
$24 - 2x = 8$ **(– 24)**
$-2x = -16$ **(÷ by -2)**
$x = 8$

Solve the following.

1 $3(2x - 7) = 9(5 + x)$

2 $4(3 - x) = -2(5x + 3)$

3 $9(3x - 2) = -5(4 - 5x)$

4 $7(4 - 3x) = 3(x + 1) - 11$

5 $12(x - 3) = 5(1 - 7x) + 6$

6 $3(6x - 1) = 2(9 - x) - 1$

7 $7 - 5(3x - 1) = -3(5 + 2x)$

8 $3(2x - 5) = 5(7 - 3x) - (6 - x)$

9 $5(3 - 2x) = 7(4x - 1) - 2(3x + 5)$

10 $9(3x - 1) - (4 - x) = 2(4 - 7x)$

ISBN: 9780170438568

Equations with fractions

Multiply **every term** by the lowest common multiple (LCM) of the denominators.

Multiply **every term** by 24 because it is LCM of 3 and 8.

Multiply **every term** by 6 because it is the LCM of 2 and 3.

Examples: 1

$$\frac{3x}{8} - 2 = \frac{x}{3}$$

$$\frac{3x}{8} \times \frac{24}{1} - 2 \times 24 = \frac{x}{3} \times \frac{24}{1}$$

$9x - 48 = 8x$ **(+ 48)**

$9x = 8x + 48$ **(– 8x)**

$x = 48$

2

$$\frac{4 + 7x}{3} = \frac{5x + 1}{2}$$

$$\frac{4 + 7x}{3} \times \frac{6}{1} = \frac{5x + 1}{2} \times \frac{6}{1}$$

$8 + 14x = 15x + 3$ **(– 15x)**

$8 - x = 3$ **(– 8)**

$-x = -5$ **(x by -1)**

$x = 5$

Solve the following.

1 $\frac{x}{5} + 2 = 9$

2 $\frac{x}{2} + \frac{2x}{5} = 9$

3 $\frac{x}{3} + \frac{x}{5} = 2$

4 $\frac{3x - 1}{2} = 7$

5 $\frac{9 - 5x}{3} = 4$

6 $\frac{3 - 4x}{4} = \frac{6x + 1}{5}$

7 $\frac{5 - 3x}{2} = \frac{7 - 2x}{5}$

8 $\frac{2x - 5}{3} - \frac{7 - x}{4} = 3$

 ISBN: 9780170438568

Forming and solving linear equations

- Define your variable using the first letter of the word whose value you need to find.
- If you need to find several values, it is usually easiest to select the smaller value as your variable.
- Often it is useful to draw a diagram.

Words used for the four basic operations:

+	plus, total, more, and, add(ed), increased by, at least, greater than	−	subtract(ed), less, decreased by, smaller than
x	of, times, multiplied by	÷	divided by, shared between

- 'Double' and 'twice' both mean multiply by 2.

Example one: Eleven people go to a movie. Adult tickets cost \$16 and children's tickets cost \$9. The total cost was \$127. Write an equation and use it to help you find the numbers of adults and children who went to the movie.

Let **a** represent the number of adults and **c** the number of children.

Always define your **variables**.

Then
$$16a + 9c = 127$$
$$16a + 9(11 - a) = 127$$

\# children = 11 – # adults

$$16a + 99 - 9a = 127$$
$$7a + 99 = 127$$
$$7a = 28$$
$$a = 4$$

∴ Four adults and seven children went to the cinema.

Always answer the question in a sentence.

Example two: Isobel is 15 years old, and she was born when her mum was 22. When will her mum be double Isobel's age? You must write an equation and use it to solve this problem.

Let **y** be the number of years until Mum is double Isobel's age.

Variable defined

Current ages: Isobel is 15 → + y years → $15 + y$; Mum is 37 → + y years → $37 + y$

Mum is double Isobel's age when:
$$2(15 + y) = 37 + y$$
$$30 + 2y = 37 + y$$
$$y = 7$$

Answer in a sentence.

∴ In seven years' time, her mum will be 44 and Isobel will be 22, i.e. her mum will be double Isobel's age.

Example three: Two sides of a rectangle are 9 cm longer than the other sides. The perimeter of the rectangle is 66 cm. Write an equation and use it to calculate the dimensions of the rectangle.

Let **s** equal the length of the shorter sides. — Variable defined

Draw a picture:

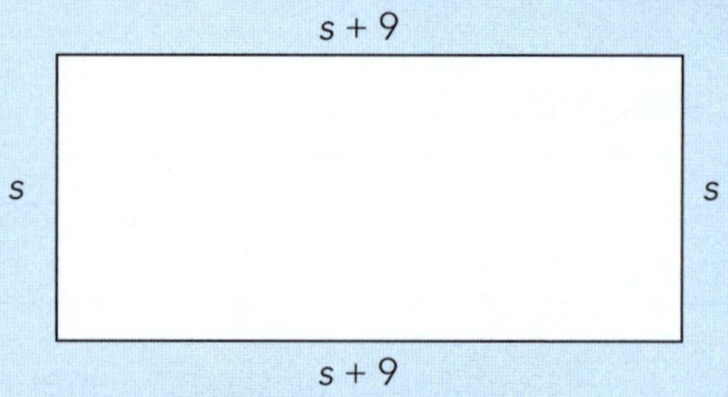

Write and equation: $s + (s + 9) + s + (s + 9) = 66$

$4s + 18 = 66$

$4s = 48$

$s = 12$ cm

∴ The shorter sides are 12 cm and the longer sides are 21 cm. — Answer in a sentence.

Write and solve equations for each of the following situations.

1 Nine people go to a show. Adult tickets cost \$15 and children's tickets cost \$7. The total cost was \$87. Write an equation and use it to help you find the numbers of adults and children who went to the show.

____

____

____

2 Anna bought nine large and six small hinges from the hardware shop. The total cost was \$43.50. The large ones cost \$1.50 more than the small ones. Write an equation and use it to help you find the price for each type of hinge.

____

____

____

3 Hine is four years old, and she was born when her mum was 22. How old will Hine be when her mum is three times her age? You must write an equation and use it to solve this problem.

____

____

____

ISBN: 9780170438568

4 Currently Dan is 20 years old, and his father is 44. How old was Dan when his father was three times his age? You must write an equation and use it to solve this problem.

5 Two sides of an isosceles triangle are 3 cm longer than the third side. Its perimeter is 39 cm. Write an equation relating the perimeter to the lengths of the sides, and use it to calculate the dimensions of the triangle.

6 Two sides of a parallelogram are 7 cm shorter than the other sides. Its perimeter is 62 cm. Write an equation relating the perimeter to the lengths of the sides, and use it to calculate the dimensions of the parallelogram.

7 An airline specifies that the maximum sum of the linear dimensions for cabin baggage (length + depth + height) is 118 cm. A bag is twice as long as it is deep, and it is 31 cm high. Write an equation relating the dimensions of the bag to the total maximum dimensions. Use it to find the maximum dimensions for this bag.

height
depth
length

8 Rua and Liz have a bag of 24 cherries. Liz keeps most for herself, but gives some to Rua. He complained, so she gave him five more. She then had one cherry fewer than 1.5 times the number of cherries that Rua had. Write an equation and solve it to find how many cherries she gave to Rua at the start.

ISBN: 9780170438568

Inequations

Inequations have **<** , **≤**, **>** or **≥** signs.

You solve an inequation in **exactly** the same way as you solve an equation **except**: if you need to **multiply or divide** the equation by a **negative number**, you must **reverse the sign**.

Examples:

1

$$5x < 20 \quad (\div 5)$$
$$x < \frac{20}{5}$$
$$x < 4$$

2

$$4x + 1 \geq 17 \quad (-1)$$
$$4x \geq 16 \quad (\div 4)$$
$$x \geq \frac{16}{4}$$
$$x \geq 4$$

3

$$3 - 7x \leq 24 \quad (-3)$$
$$-7x \leq 21 \quad (\div \text{ by } -7)$$
$$x \geq -3$$

Dividing by a **negative** number ⇒ reverse the sign.

4

$$\frac{7-x}{5} > \frac{2x-1}{3} \quad (\times \text{ by } 15)$$
$$21 - 3x > 10x - 5 \quad (-21)$$
$$-3x > 10x - 26 \quad (-10x)$$
$$-13x > -26 \quad (\div \text{ by } -13)$$
$$x < 2$$

Solve the following.

1 $3x > 24$

2 $2x - 9 < 13$

3 $19 - 4x \leq 3$

4 $2x - 9 > 5x + 12$

5 $2(3 - 5x) < -4$

6 $3(2x - 5) \leq 7(4 + x)$

7 $1 - \frac{6x}{11} < 4$

8 $\frac{5-3x}{2} \geq \frac{7-2x}{5}$

 ISBN: 9780170438568

Forming and solving inequations

Your answers MUST have one of the following signs: >, <, ≥ or ≤, NOT an = sign.

Write and solve inequations for each of the following situations.

1 Ali writes a puzzle. Eleven minus four times a mystery number must be greater than one. Write an inequation and solve it to find what values the mystery number can take.

2 The length of a rectangular cake is one and a half times its width. The length of a decoration that will wrap around the perimeter of the cake is 75 cm. Write an inequation and solve it to find the maximum dimensions of the cake.

3 Mick, Adam and Sam went out for dinner again. They spent less than $60 altogether. This time Adam's and Mick's meals cost the same, but Sam's meal cost $3 less than Adam's. Write an inequation for the total cost of the dinner in terms of M, the cost of Mick's meal. Use it to calculate the maximum costs for each meal.

4 Ruth has a budget of $100 for her birthday party. She needs at least $35 for food and will use the remainder for movie tickets for herself and her friends. These cost $8.50 each. Write an inequation and solve it to find the maximum number of friends that she can invite.

5 Penny has at least $4.50 plus $\frac{2}{5}$ of the amount of money that Suzie has. Write an expression for the amount Penny has (p) in terms of how much Suzie has (s). If Suzie has $11.50, use your expression to calculate how much Penny has.

ISBN: 9780170438568

Equations with powers

- Isolate the term with the variable on the left.
- Then write both sides so they are raised to the same power.

Examples: Solve the following.

1
$x^3 = -125$
$x^3 = (-5)^3$
$x = -5$

2
$x^3 - 19 = 8$ **(+ 19)**
$x^3 = 27$
$x^3 = 3^3$
$x = 3$

3
$2x^4 = 32$ **(÷ by 2)**
$x^4 = 16$
$x^4 = 2^4$
$x = 2$

4
$\frac{1}{4}x^5 = -8$ **(x by 4)**
$x^5 = -32$
$x^5 = (-2)^5$
$x = -2$

Solve the following.

1 $x^2 = 81$

2 $x^3 = 27$

3 $x^6 = 64$

4 $x^3 = -64$

5 $x^3 + 1 = 9$

6 $x^2 - 13 = 36$

7 $2x^3 = -54$

8 $3x^4 = 48$

9 $\frac{1}{2}x^2 = 18$

10 $\frac{1}{4}x^3 + 1 = -15$

ISBN: 9780170438568

Quadratic expressions

- An expression in which the highest power of the variable is 2 is known as a quadratic expression.
- Usually these contain an x^2, but in factorised form quadratic expressions may look like $(x \pm a)(x \pm b)$ or $x(x \pm a)$.
- When graphed, quadratics form curves known as parabolas:

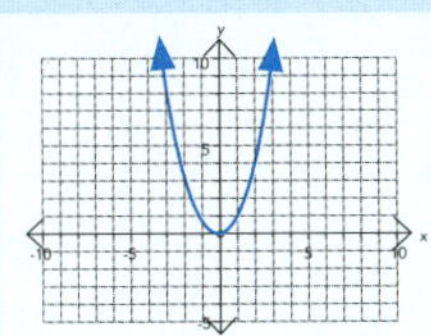

or

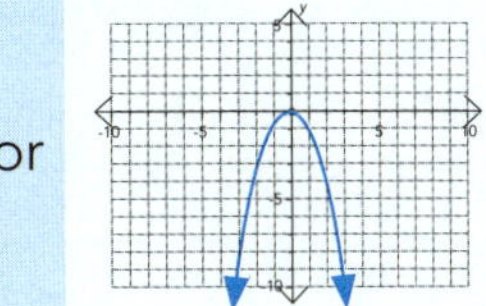

Expanding quadratic expressions

Remember **FOIL:**

Multiply the **F**irsts
Outers
Inners
Lasts

F O I L

e.g. $(x + 2)(3x - 5) = 3x^2 - 5x + 6x - 10$
$= 3x^2 + x - 10$

Combine like terms

Examples:

1 $(x + 4)(x + 5) = x^2 + 5x + 4x + 20$
$= x^2 + 9x + 20$

2 $(5x + 3)(x - 2) = 5x^2 - 10x + 3x - 6$
$= 5x^2 - 7x - 6$

3 $(x + 7)^2 = (x + 7)(x + 7)$
$= x^2 + 7x + 7x + 49$
$= x^2 + 14x + 49$

4 $(4x - 2)^2 = (4x - 2)(4x - 2)$
$= 16x^2 - 8x - 8x + 4$
$= 16x^2 - 16x + 4$

5 $(x + 4)(x - 4) = x^2 - 4x + 4x - 16$
$= x^2 - 16$

Expand and simplify these.

1 $(x + 3)(x + 4)$

____

____

2 $(x + 7)(x - 2)$

____

____

ISBN: 9780170438568

3 $(x + 3)(x - 9)$

4 $(x - 3)(x - 10)$

5 $(5 + x)(x + 8)$

6 $(1 + 3x)(x - 7)$

7 $(x + 8)^2$

8 $(x - 5)^2$

9 $(3x + 1)(x - 2)$

10 $(3x - 1)(2x + 6)$

11 $(3 - x)^2$

12 $(3x - 5)^2$

13 $3x(2x - 5)$

14 $(x + 1)(x - 1)$

15 $(5x - 2)(4x + 3)$

16 $(7 - 2x)(4 + 3x)$

17 $(4x - 3)(4x + 3)$

18 $(5 - 2x)(5 + 2x)$

ISBN: 9780170438568

Factorising quadratic expressions

1 Where the coefficient of x^2 is 1

Steps:

1 List all the factors of the constant: $x^2 + 3x - 40$

1, 40
2, 20
4, 10
5, 8

2 Select the pair that could add or subtract to give the coefficient of x: $x^2 + 3x - 40$

$-5 + 8 = +3$

3 The factors are $(x - 5)(x + 8)$

Check your answer by expanding the brackets using **FOIL** — you should get the original expression: $(x - 5)(x + 8) = x^2 + 3x - 40$

Examples:

1 $x^2 + 7x + 12 = (x + 3)(x + 4)$

1, 12
2, 6
3, 4

2 $x^2 - 11x + 30 = (x - 5)(x - 6)$

1, 30
2, 15
3, 10
-5, -6

3 $x^2 - 7x - 18 = (x + 2)(x - 9)$

1, 18
+2, -9
3, 6

4 $x^2 - 36 = x^2 + 0x - 36 = (x + 6)(x - 6)$

1, 36
2, 18
3, 12
4, 9
+6, -6

Insert a 'fake' x term.

Factorise the following.

1 $x^2 + 6x + 8$

____

____

2 $x^2 + 8x + 7$

____

____

ISBN: 9780170438568

3 $x^2 + 13x + 36$

4 $x^2 + 4x - 12$

5 $x^2 + 19x + 60$

6 $x^2 - 9x + 18$

7 $x^2 - 2x - 24$

8 $x^2 - 10x + 25$

9 $x^2 - 5x - 6$

10 $x^2 - x - 6$

11 $x^2 - 5x + 6$

12 $x^2 + 5x - 6$

13 $x^2 - 25$

14 $49 - x^2$

15 $9 - 16x^2$

16 $25 - 9x^2$

17 $24 + 11x + x^2$

18 $15 - 8x + x^2$

ISBN: 9780170438568

2 Where the coefficient of x^2 is not 1, but there is a common factor

Steps:

2 divides into the coefficients and the constant.

1 Take out the common factor first: $2x^2 + 4x - 6 = 2(x^2 + 2x - 3)$

2 Factorise the contents of the bracket: $= 2(x + 3)(x - 1)$

Examples:

1 $6x^2 + 18x - 60 = 6(x^2 + 3x - 10)$
$= 6(x + 5)(x - 2)$

2 $2x^2 + 20x + 50 = 2(x^2 + 10x + 25)$
$= 2(x + 5)(x + 5)$
$= 2(x + 5)^2$

3 $4x^2 - 36 = 4(x^2 - 9)$
$= 4(x - 3)(x + 3)$

Factorise the following.

1 $4x^2 + 32x + 48$

2 $2x^2 + 14x + 24$

3 $5x^2 - 5x - 30$

4 $3x^2 + 15x - 42$

5 $10x^2 - 40x + 40$

6 $5x^2 - 180$

7 $5x^2 - 10x + 5$

8 $8x^2 - 8$

9 $2x^2 - 4x - 6$

10 $200 - 8x^2$

ISBN: 9780170438568

3 Where the coefficient of x^2 is not 1, and there is no common factor

Example: Factorise $10x^2 + 3x - 4$

Step 1: Multiply the coefficient of x^2 by the constant.

$10 \times -4 = -40$

Step 2: List factors of 40:
1, 40
2, 20
4, 10
5, 8

Step 3: Select the pair that can add to produce the coefficient of x.

$-5 + 8 = 3$

Step 4: Rewrite the quadratic with this pair as coefficients of **two** x terms.

$10x^2 - 5x + 8x - 4$

Step 5: Factorise the first pair of terms and the second pair of terms.

$5x(2x - 1) + 4(2x - 1)$

The contents of both brackets should be the same!

Step 6: Factorise, making the bracket contents the common factor.

$(2x - 1)(5x + 4)$

Examples showing what you should get at each step:

1 $18x^2 - 9x - 5$

Step 1: $18 \times -5 = -90$

Step 2:
1, 90
2, 45
3, 30
5, 18
6, 15
9, 10

Step 3: $+6 - 15 = -9$

Step 4: $18x^2 + 6x - 15x - 5$

Step 5: $6x(3x + 1) - 5(3x + 1)$

Step 6: $(3x + 1)(6x - 5)$

2 $10x^2 - 19x + 6$

Step 1: $10 \times 6 = 60$

Step 2:
1, 60
2, 30
3, 20
4, 15
5, 12
6, 10

Step 3: $-4 - 15 = -19$

Step 4: $10x^2 - 4x - 15x + 6$

Step 5: $2x(5x - 2) - 3(5x - 2)$

Step 6: $(5x - 2)(2x - 3)$

Signs!

Factorise the following.

1 $2x^2 + 11x + 12$

2 $2x^2 + 13x + 6$

 ISBN: 9780170438568

3 $4x^2 + 16x + 15$

4 $2x^2 - 5x - 12$

5 $3x^2 - 20x + 12$

6 $2x^2 - 5x + 3$

7 $4x^2 - 15x - 25$

8 $6x^2 - 7x - 3$

9 $10x^2 + 13x - 3$

10 $9x^2 - 9x + 2$

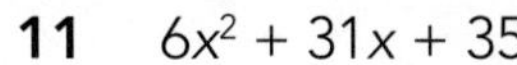

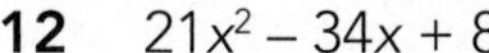

11 $6x^2 + 31x + 35$

12 $21x^2 - 34x + 8$

ISBN: 9780170438568

Simplifying quadratic fractions

Factorise the top and bottom, then cancel the common factors.

Examples:

1 $\dfrac{x^2 + 7x + 12}{x + 3} = \dfrac{(x + 4)(x + 3)}{x + 3}$

$= (x + 4)$

2 $\dfrac{x^2 + 4x + 4}{x^2 - x - 6} = \dfrac{(x + 2)(x + 2)}{(x + 2)(x - 3)}$

$= \dfrac{x + 2}{x - 3}$

Simplfy these.

1 $\dfrac{2x + 8}{x + 4}$

2 $\dfrac{x - 1}{4x - 4}$

3 $\dfrac{x^2 + 3x}{x + 3}$

4 $\dfrac{2x - 12}{3x - 18}$

5 $\dfrac{x^2 + 10x + 16}{x + 2}$

6 $\dfrac{x^2 - 7x + 12}{x - 3}$

7 $\dfrac{x + 4}{x^2 + x - 12}$

8 $\dfrac{3x - 9}{x^2 + 2x - 15}$

9 $\dfrac{x^2 + 4x + 3}{x^2 + 2x - 3}$

10 $\dfrac{x^2 + 4x + 4}{x^2 + 6x + 8}$

ISBN: 9780170438568

Exam questions

Answer the following questions. In all cases, simplify your answers where possible.

1 The lengths of the sides of a rectangle are $6x - 5$ and $2x + 7$. Write an expanded expression for its area.

2 If $(3x - 2)(5x - 4) = ax^2 + bx + c$, write down the values for a, b and c.

3 The area of a rectangle can be expressed as $14x^2 - 3x - 2$. State the length and the width of this rectangle.

Length = ________ Width = ________

4 $a \times b = 5x^2 + 18x - 8$. Write expressions for a and b.

5 The area of a rectangle can be expressed as $16x^2 - 25$. Give expressions for the length and the width of this rectangle.

Length = ________ Width = ________

6 The area of a square can be expressed as $25x^2 - 20x + 4$. Give an expression for the length of one side of this square.

Length = ________

7 A right-angled triangle's area is given by $21x^2 - 9x$. The length of its base is $6x$. How high is the triangle?

Length = ________

8 $8x^2 - 2x - 15 = (ax + b)(cx + d)$. Find integer values for a, b, c and d.

Solving quadratic equations

1 Factorised quadratic equations

If $a \times b = 0$

then either $a = 0$ or $b = 0$

Remember, the x sign has not been written in here.

So, if we have a quadratic equation $(x - 2)(x + 3) = 0$

then either $(x - 2) = 0$ or $(x + 3) = 0$

So either $x = 2$ or $x = -3$

Note:

- Most quadratic equations that you come across will have two solutions. However, some will have just one solution, and some will have no real solutions.
- There may be two solutions, but in a practical situation, one may not work.

Examples:

1

$(x + 7)(x - 2) = 0$

Either $(x + 7) = 0$

$x = -7$

or $(x - 2) = 0$

$x = 2$

2

$(x + 3)^2 = 0$

$\therefore (x + 3) = 0$

$x = -3$

One solution only

3

$4x(x - 13) = 0$

Clearly $4 \neq 0$

So either $x = 0$

or $(x - 13) = 0$

$x = 13$

4

$(2x - 1)(3x + 4) = 0$

Either $(2x - 1) = 0$

$2x = 1$

$x = \frac{1}{2}$

or $(3x + 4) = 0$

$3x = -4$

$x = -\frac{3}{4}$

Solve the following equations.

1 $(x + 9)(x - 1) = 0$

____

____

2 $(x - 7)(x + 2) = 0$

____

____

3 $(x + 1)(x + 4) = 0$

____

____

4 $(x - 3)(x - 8) = 0$

____

____

ISBN: 9780170438568

5 $(x + 7)(x - 3) = 0$

6 $(x - 3)^2 = 0$

7 $(x + 6)^2 = 0$

8 $(x + \frac{1}{2})^2 = 0$

9 $(x + 2)(5 - x) = 0$

10 $(x - 3)(4 + x) = 0$

11 $x(x - 1) = 0$

12 $5x(x + 9) = 0$

13 $-x(3x - 1) = 0$

14 $7x(2x + 5) = 0$

15 $-5x(x - 1) = 0$

16 $\frac{1}{2}x(x - \frac{2}{3}) = 0$

ISBN: 9780170438568

2 Unfactorised quadratic equations

Do this by factorising the equation first.

Example: $x^2 - 3x - 10 = 0$

Step 1: Factorise: $(x + 2)(x - 5) = 0$

Step 2: Solve by considering what happens when each bracket equals 0.

Either $(x + 2) = 0$

$\mathbf{x = -2}$

or $(x - 5) = 0$

$\mathbf{x = 5}$

So, if $x^2 - 3x - 10 = 0$, then x is either -2 or 5.

Examples:

1
$$x^2 + 5x + 4 = 0$$
$$(x + 4)(x + 1) = 0$$
Either $(x + 4) = 0$
$$x = -4$$
or $(x + 1) = 0$
$$x = -1$$

2
$$x^2 + 10x + 25 = 0$$
$$(x + 5)(x + 5) = 0$$
$$\therefore (x + 5) = 0$$
$$x = -5$$

One solution only

3
$$x^2 - 25 = 0$$
$$(x + 5)(x - 5) = 0$$
Either $(x + 5) = 0$
$$x = -5$$
or $(x - 5) = 0$
$$x = 5$$

4
$$6x^2 + 5x - 6 = 0$$
$$(2x + 3)(3x - 2) = 0$$
Either $(2x + 3) = 0$
$$x = -\frac{3}{2}$$
or $(3x - 2) = 0$
$$x = \frac{2}{3}$$

Solve the following equations.

1 $x^2 + 5x + 6 = 0$

____

____

2 $x^2 - 3x - 10 = 0$

____

____

3 $x^2 - 10x + 24 = 0$

____

____

4 $x^2 - 9x = 0$

____

____

 ISBN: 9780170438568

5 $x^2 + 3x = 0$

6 $x^2 - 4x - 21 = 0$

7 $x^2 + 14x + 49 = 0$

8 $x^2 + 3x - 28 = 0$

9 $x^2 - 6x + 9 = 0$

10 $x^2 - 5x - 24 = 0$

11 $x^2 + 10x + 25 = 0$

12 $x^2 - 4 = 0$

13 $x^2 - 81 = 0$

14 $x^2 - 12x + 36 = 0$

15 $x^2 - 11x - 60 = 0$

16 $x^2 - 11x + 24 = 0$

17 $3x^2 + 10x - 8 = 0$

18 $5x^2 - 33x - 14 = 0$

19 $2x^2 + 3x - 2 = 0$

20 $15x^2 + 7x - 2 = 0$

ISBN: 9780170438568

3 Quadratic equations where rearrangement is needed

- Sometimes quadratic equations are not organised like the ones in the previous exercise.
- Use the equation-solving rules to reorganise them into an $ax^2 + bx + c = 0$ form.
- Then solve as usual.

Examples:

1
$$x^2 - 7x = -6 \quad \textbf{(+ 6)}$$
$$x^2 - 7x + 6 = 0$$
$$(x - 6)(x - 1) = 0$$
$$x = 6 \text{ or } 1$$

2
$$5x + 4 = -x^2 \quad \textbf{(+ }\boldsymbol{x^2}\textbf{)}$$
$$x^2 + 5x + 4 = 0$$
$$(x + 4)(x + 1) = 0$$
$$x = -4 \text{ or } -1$$

3
$$-2x = x^2 - 8 \quad \textbf{(– 2x)}$$
$$x^2 + 2x - 8 = 0$$
$$(x + 4)(x - 2) = 0$$
$$x = -4 \text{ or } 2$$

4
$$x^2 - 3x - 6 = 4 \quad \textbf{(– 4)}$$
$$x^2 - 3x - 10 = 0$$
$$(x + 2)(x - 5) = 0$$
$$x = -2 \text{ or } 5$$

5
$$(x - 6)(x + 1) = 8$$
$$x^2 - 5x - 6 = 8 \quad \textbf{(– 8)}$$
$$x^2 - 5x - 14 = 0$$
$$(x + 2)(x - 7) = 0$$
$$x = -2 \text{ or } 7$$

6
$$\frac{18 - 7x}{x} = x \quad \textbf{(x by x)}$$
$$18 - 7x = x^2 \quad \textbf{(– 18, + 7x)}$$
$$x^2 + 7x - 18 = 0$$
$$(x + 9)(x - 2) = 0$$
$$x = -9 \text{ or } 2$$

Solve the following equations.

1 $x^2 - x = 90$

2 $x^2 + 27 = 12x$

3 $14 - 5x = x^2$

4 $12 = 7x - x^2$

5 $x(x - 8) = 48$

6 $x = \dfrac{36 - 5x}{x}$

7 $(x - 3)(x - 8) = 84$

8 $2x^2 + 35 = x(x + 12)$

ISBN: 9780170438568

Forming and solving quadratic equations

Hints:
- Read what the question asks for. Call this x.
- If there are two things to find, call the smaller one x and the bigger one $(x + ?)$.
- Usually you will get two answers. Test both to see if they work. If you have to reject one answer because it doesn't work, explain why.

1 Problems about numbers and ages

Finding unknown numbers
- Even numbers: call these $2x$, e.g. 10, where $x = 5$.
- Odd numbers: call these $2x + 1$, e.g. 15, where $x = 7$.
- Consecutive numbers come after each other: call these x, $x + 1$, $x + 2$, e.g. 6, 7, 8, where $x = 6$.
- Consecutive even numbers: call these $2x$, $2x + 2$, $2x + 4$, e.g. 6, 8, 10, where $x = 3$.
- Consecutive odd numbers: call these $2x + 1$, $2x + 3$, $2x + 5$, e.g. 9, 11, 13, where $x = 4$.
- Age questions: give the youngest the age x.

Example: Mark thinks of a positive number. He subtracts 3, squares the result, then adds 2. His answer is 27. What number was he thinking of?

Call the number he first thought of x.

subtracts 3, $x - 3$

squares the result, $(x - 3)^2$

adds 2, $(x - 3)^2 + 2$

$$(x - 3)^2 + 2 = 27$$
$$x^2 - 6x + 9 + 2 = 27$$
$$x^2 - 6x - 16 = 0$$
$$(x - 8)(x + 2) = 0$$
$$\therefore\ x = 8 \text{ or } x = -2$$

His number had to be positive, so -2 is not a possibility; so he thought of 8.

Try for yourself.

Mariama thinks of a positive number. She squares it, then adds 15 times the number she thought of. Her answer is 100. What number was she thinking of?

ISBN: 9780170438568

Write and solve quadratic equations to find the unknown numbers.

1 Ben thinks of a number. He adds 4 to it, then squares the result, then subtracts 5. His answer is 31. What numbers could he have been thinking of?

2 Ellie thinks of a number. She squares it, then multiplies the answer by 4. Her answer is 100. What numbers could she be thinking of?

3 Matt thinks of a positive even number. He multiplies it by the next (bigger) even number. His answer is 168. What positive even number was he thinking of?

4 Alex squares a positive number, subtracts 6 times the number and gets an answer of 16. What number did he start with?

5 Boyd thinks of a positive number. He multiplies it by 5 more than the number. The answer is 84. What number did he think of?

6 Petra squared a positive number. She got the same answer if she multiplied the number by 5 and then added 36. What number did she start with?

ISBN: 9780170438568

7 Miranda thinks of two positive numbers. One is 8 more than the other. The product of the two numbers equals 3 times their sum. What are the numbers she thought of?

8 Amber is 3 years younger than her sister. In 4 years the product of their ages will be 238. How old is Amber?

9 Peter was born when his mother was 22. Ten years ago the product of their ages was 135. How old is Peter now?

10 Jacob's dad is 24 years older than he is. Nine years ago the product of their ages was 145. How old are they now?

11 Liam is the youngest of three brothers who were born at two-year intervals. The product of the ages of the two younger brothers is 52 less than the product of the ages of the two older brothers. Calculate the ages of the three brothers.

12 Wiremu has some lollies. His brother has 5 more than he has. They are both given another 3 lollies, and then the product of the numbers of lollies they each have is 150. How many lollies did Wiremu start with?

ISBN: 9780170438568

2 Problems about lengths, areas and volumes

Hint: If you are not given one, do your own sketch and add all the information to it.

Example: An isosceles triangle has both a base and a vertical height of $3x$ cm. A similar but smaller isosceles triangle is cut out of the middle of the big triangle. It has both a base and a height of x cm. The area of the remaining shape is 196 cm^2. Calculate the value of x, and the base of the large triangle.

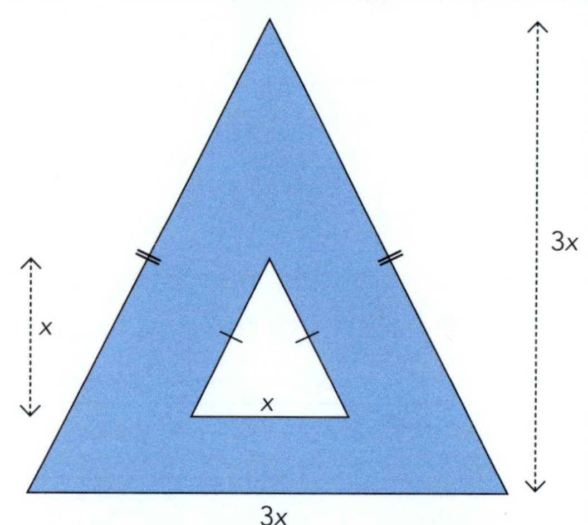

$$\text{Area big } \Delta = \frac{1}{2}bh$$
$$= \frac{1}{2}(3x)^2$$
$$= \frac{9x^2}{2}$$
$$\text{Area small } \Delta = \frac{1}{2}bh$$
$$= \frac{1}{2}(x)^2$$
$$= \frac{x^2}{2}$$
$$\therefore \text{ Shaded area} = \frac{9x^2}{2} - \frac{x^2}{2} = 4x^2 = 196$$
$$x^2 = 49$$
$$x = \pm 7$$

A length has to be positive, so -7 is not a possibility; so $x = 7$.
The base of the large triangle is 21 cm wide.

Always answer the question in a **sentence**.

Try for yourself.

A rectangle is $5x$ cm wide and $4x$ cm high. A smaller rectangular shape is cut from this rectangle. This smaller rectangle is $3x$ cm wide and $2x$ cm high. The remaining area is 350 cm^2. Calculate the value of x, and the dimensions of the big rectangle.

ISBN: 9780170438568

Solve quadratic equations to find the unknowns.

1 The top and bottom of a rectangle are 5 cm longer than its sides. The area of a rectangle is 204 cm^2. Calculate the lengths of the sides.

2 A cuboidal box has a base that is $2y$ cm wide and y cm deep. The box is 5 cm high. If its volume is 360 cm^3, calculate its dimensions.

3 Another cuboidal box has a base that is $3y$ cm wide and $(y + 2)$ cm deep. The box is 10 cm high. If its volume is 1050 cm^3, calculate its dimensions.

4 The base of a triangle is $2z$ cm, and its perpendicular height is $(z + 4)$ cm. Its area is 45 cm^2. Calculate the length of the base.

5 A large 'L' is to be part of a sign on a building. The width of each arm of the 'L' is x metres. If the area of the 'L' is 45 m^2, calculate the width of each arm.

x
8
x
x + 4

ISBN: 9780170438568

6 A letter 'H' is put on another building. The width of each arm of the 'H' is x metres, and the shape is symmetrical. If the area of the 'H' is 128 m^2, calculate the width of each arm.

7 The length of the short parallel side of a trapezium is the same as its height. The long parallel side is 8 cm longer than the short parallel side. The area of the trapezium is 45 cm^2. What is the height of the trapezium?

8 The sides of a right-angled triangle are x cm, $(x + 7)$ cm and $(x + 9)$ cm. Use the Theorem of Pythagoras to calculate the lengths of the three sides.

9 A rectangular piece of fabric has a square cut out of one corner. The longer edge of the piece of fabric is 6 cm longer than the side of the square, and the shorter edge is 3 cm longer than the side of the square. If the remaining area is 54 cm^2, calculate the dimensions of the original rectangular piece of fabric.

10 The floors of a square living room and a rectangular bedroom are to be covered using a total of 273 m^2 of carpet. One side of the bedroom is 7 m shorter than the side of the living room, and the other side is 9 m shorter than the living room. Calculate the length of the living room.

ISBN: 9780170438568

3 Challenges

Examples:

1 Solve $x^2 - ax + 28 = 4$, where a is a positive number. If the difference between the solutions to the equation is 10, find the value of a, and the possible solutions to the equation.

Step 1: Reorganise to give 0 on one side: $x^2 - ax + 24 = 0$

Step 2: List the factors of 24:
1, 24
2, 12
3, 8
4, 6

Both pairs could give solutions with a difference of 10.

Step 3: Decide which pair of numbers to use: The '+ 24' in Step 1 $\Rightarrow$ solutions must have the same sign.
$\therefore$ Use 2, 12.

Step 4: Decide which pair of factors to use: $(x + 12)(x + 2)$ or $(x - 12)(x - 2)$
Because a must be positive, we must use the second pair.

$(x - 12)(x - 2) = x^2 - 14x + 24$

$\therefore a = 14$ and solutions to the equation are $x = 2$ or 12.

2 For what values will p be positive for the following relationship? $p = n^2 + n - 12$
This is the same as being asked to solve $n^2 + n - 12 > 0$ so:

Step 1: Solve $n^2 + n - 12 = 0$: $(n + 4)(n - 3) = 0$
$\therefore n = -4$ or $n = 3$

Step 2: Decide the range of values for which $n^2 + n - 12 > 0$.

Method 1: Sketch the graph:
$\therefore p > 0$ if $n < -4$ or $n > 3$

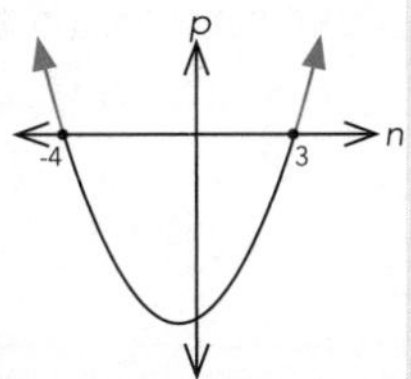

Method 2: p must be positive either for $n < -4$ and $n > 3$, or for $-4 < n < 3$.
Try substituting a point that is clearly in one region or the other.
e.g. Let $n = 0$ (easy to substitute and clearly within the region $-4 < n < 3$)

$0^2 + 0 - 12 = -12$ — $\therefore$ p would **not** be positive

So p must be positive for $n < -4$ and $n > 3$.

3 Consider the sequence $T = 2n^2 + n - 3$. Give the rule for finding the **difference** between consecutive terms.
Hint: Consecutive terms follow each other, e.g. the 4th and 5th terms, or the 10th and 11th terms, or the nth and $(n + 1)$th term.

The $(n + 1)$th term – the nth term
$$= [2(n + 1)^2 + (n + 1) - 3] - [2n^2 + n - 3]$$
$$= [2(n^2 + 2n + 1) + (n + 1) - 3] - [2n^2 + n - 3]$$
$$= 2n^2 + 4n + 2 + n + 1 - 3 - 2n^2 - n + 3$$
$$= 4n + 3$$

ISBN: 9780170438568

Solve the following questions.

1 $x^2 - ax + 26 = 8$, where a is a positive number. If the difference between the solutions to the equation is 7, find the value of a, and the possible solutions to the equation.

2 $x^2 - ax - 20 = -8$, where a is a positive number. If the difference between the solutions to the equation is 8, find the value of a, and the possible solutions to the equation.

3 $x^2 - ax - 20 = -8$, where a is a positive number. If the difference between the solutions to the equation is 7, find the value of a, and the possible solutions to the equation.

4 $x^2 + ax + 20 = 14$, where a is a positive number. If the difference between the solutions to the equation is 5, find the value of a, and the possible solutions to the equation.

5 $x^2 - ax + 25 = 9$, where a is a positive number. If there is only one solution to this equation, find the value of a, and the solution to the equation.

6 For what values will p be positive for the following relationship? $p = n^2 - 8n + 15$.

  ISBN: 9780170438568

7 For what values will p be negative for the following relationship? $p = n^2 - 5n - 24$

8 For what values will p be positive for the following relationship? $p = n^2 + n - 12$

9 For what values will p be negative for the following relationship? $p = 20 - 12n + n^2$

10 Consider the sequence $T = 2n^2 - 3n + 4$. Give the rule for finding the **difference** between consecutive terms.

11 The sequence for triangular numbers is given by $T = \frac{1}{2}(n^2 + n)$. Give the rule for finding the **difference** between consecutive triangular numbers.

12 The sequence for hexagonal numbers is given by $T = 3n^2 - 3n + 1$. Give the rule for finding the **difference** between consecutive hexagonal numbers.

ISBN: 9780170438568

Exponential equations

Basic exponential equations

- In exponential equations, the unknown is the exponent (power).
- **Remember:**

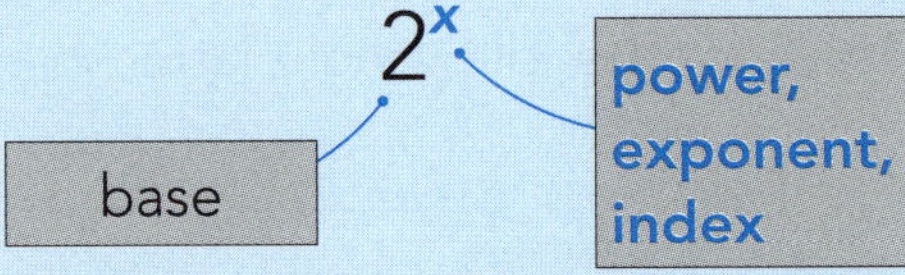

- The only way for you to solve these is by listing the powers of the base.

Examples: **1** $2^x = 16$

List the powers of 2: $2^0 = 1$, $2^1 = 2$, $2^2 = 4$, $2^3 = 8$, $2^4 = 16$

$\therefore\ 2^x = 2^4$

So $x = 4$

2 $3^x = 243$

List the powers of 3: $3^0 = 1$, $3^1 = 3$, $3^2 = 9$, $3^3 = 27$, $3^4 = 81$, $3^5 = 243$

$\therefore\ 3^x = 3^5$

So $x = 5$

Solve the following exponential equations.

1 $2^x = 8$

2 $3^x = 81$

3 $5^x = 125$

4 $4^x = 256$

5 $7^x = 7$

6 $6^x = 1$

7 $10^x = 1\ 000\ 000$

8 $0.1^x = 0.0001$

 ISBN: 9780170438568

Using equation rules in exponential equations

Normal rules for solving equations can be used for exponential equations.

Examples: **1**

$5 \times 2^x = 40$ **(÷ by 5)**

$2^x = 8$

$x = 3$

2

$3^{(x-1)} = 81$

$3^4 = 81$

$\therefore (x - 1) = 4$ **(+ 1)**

$x = 5$

Solve the following exponential equations.

1 $3 \times 2^x = 48$

2 $5^{(x+1)} = 125$

3 $3^{(2x)} = 81$

4 $5 \times 4^x = 320$

5 $3 \times 10^x = 30\,000$

6 $7^{(x+2)} = 49$

7 $2 \times 5^{(x+1)} = 250$

8 $5 \times 10^{(x-2)} = 50\,000$

9 $\frac{1}{2} \times 2^{(x-3)} = 32$

10 $4 \times 3^{(2x+1)} = 972$

11 $3 \times 10^{(2x-3)} = 300\,000$

12 $64 \times \left(\frac{1}{2}\right)^{2(x+2)} = 1$

ISBN: 9780170438568

Simultaneous equations

- 'Simultaneous' means 'at the same time'.
- For simultaneous equations, you need to be able to solve two equations that are both true 'at the same time'.
- Because there are two equations, there are also two variables.
- There are two methods for solving these, and which one you use depends on how the equations are structured.

1 Substitution

- Substitution is easiest when one of the equations is expressed as **x = ...** or **y = ...**
- As the name suggests, you simply substitute the **x = ...** into the other equation.
- You should number each equation, and say what you are doing at each step.

Examples:

1 Solve the equations $x = 2y - 1$ and $3x + y = 18$

Number the equations.

$x = 2y - 1$ ①

$3x + y = 18$ ②

Substitute ① into ②: $3(2y - 1) + y = 18$

From equation ① we know that **x** and **2y − 1** are equal.

$6y - 3 + y = 18$

$7y - 3 = 18$

$7y = 21$

$\mathbf{y = 3}$

Substitute for y in ①: $x = 2(3) - 1$

$\mathbf{x = 5}$

2 Solve the equations $5y - 2x = 22$ and $y = x - 1$

$5y - 2x = 22$ ①

$y = x - 1$ ②

Substitute ② into ①: $5(x - 1) - 2x = 22$

Select ② because it has the form $y =$ .....

$5x - 5 - 2x = 22$

$3x - 5 = 22$

$3x = 27$

$\mathbf{x = 9}$

Substitute for x in ②: $y = 9 - 1$

Select whichever equation is easier for substitution.

$\mathbf{y = 8}$

ISBN: 9780170438568

Solve the following simultaneous equations.

1
$$x = y + 1$$
$$x + 3y = 21$$

2
$$y = 2x + 1$$
$$x + 2y = 27$$

3
$$x = 3y - 1$$
$$5x - y = 23$$

4
$$4x + y = 40$$
$$y = 3x - 2$$

5
$$y = 3x - 5$$
$$2y - x = 90$$

6
$$3x + 2y = 13$$
$$y = 5 - 2x$$

7
$$x = 1 - y$$
$$2x - 7y = 20$$

8
$$x = 3y$$
$$15 + 2x = y$$

9
$$x = 4 - 3y$$
$$3x - 5y - 54 = 0$$

10
$$11x - 3y + 34 = 0$$
$$y = 7x - 2$$

ISBN: 9780170438568

2 Elimination

- Elimination is easiest when the two equations have the same structure.

For example:

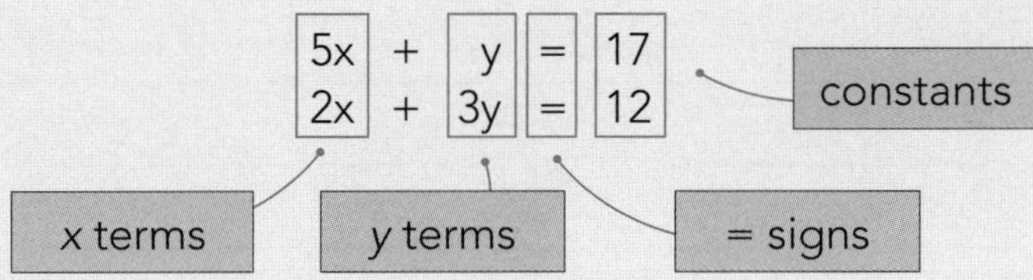

- You solve these by multiplying one or both equations by a constant in order to make either the x terms or the y terms the **same size** but with **different signs**.
- Then **add** the two equations to produce an equation with one variable only.
- You should number each equation, and say what you are doing at each step.

Examples:

1 Solve the equations $5x + y = 17$ and $2x + 3y = 12$.

	$5x + y = 17$	①
	$2x + 3y = 12$	②
Multiply ① by -3:	$-15x - 3y = -51$	③
Add ② and ③:	$-13x = -39$	
	$\therefore$ **$x = 3$**	
Substitute for x in ①:	$5(3) + y = 17$	
	$\therefore$ **$y = 2$**	

Call this ③ because it is a new equation.

2 Solve the equations $5x - 2y = 27$ and $4x - 3y = 16$.

	$5x - 2y = 27$	①
	$4x - 3y = 16$	②
Multiply ① by -3	$-15x + 6y = -81$	③
Multiply ② by 2:	$8x - 6y = 32$	④
Add ③ and ④:	$-7x = -49$	
	$\therefore$ **$x = 7$**	
Substitute for x in ①:	$5(7) - 2y = 27$	
	$-2y = 27 - 35$	
	$\therefore$ **$y = 4$**	

You need to multiply both equations by a constant.

Solve the following simultaneous equations.

1 $4x + y = 19$
$2x - y = 5$

____

____

____

____

____

2 $5x + 3y = 25$
$2x + 3y = 19$

____

____

____

____

____

ISBN: 9780170438568

3

$$x - 2y = 21$$
$$3x + y = 28$$

4

$$5x + 2y = 24$$
$$x + 3y = 23$$

5

$$2y - 3x = 13$$
$$y - 4x = 9$$

6

$$5x + 2y = 0$$
$$-x + 6y = 32$$

7

$$2x + 3y = 18$$
$$5x + 2y = 23$$

8

$$4x - 3y = 25$$
$$5x + 6y = 41$$

9

$$3x + 2y = 3$$
$$7x - 3y = -39$$

10

$$8x + 3y = -4$$
$$5x + 4y = 6$$

11

$$12x + 7y = 3$$
$$5x - 4y = 22$$

12

$$-2x + 9y = 4$$
$$-7x + 12y = 53$$

ISBN: 9780170438568

Forming and solving simultaneous equations

Hint: Read what the question asks for. Define your variables using their first letters.

Examples:

1 Tom had twice as many marbles as Emily had. If Emily had another 25 marbles, she would have had three times as many as Tom. How many marbles did each person have?

Call the number of marbles Tom has **t**.
Call the number of marbles Emily has **e**.

Always define your **variables**.

$t = 2e$ ①
$e + 25 = 3t$ ②
Substitute ① into ②: $e + 25 = 3(2e)$
$e + 25 = 6e$ $(-e)$
$25 = 5e$
$\mathbf{e = 5}$
Substitute for e in ①: $t = 2(5)$
$\mathbf{t = 10}$

∴ **Tom had 10 marbles and Emily had 5.**

Always answer the question in a **sentence**.

2 Nick buys 6 peaches and 5 apples and they cost him \$5.70.
Four peaches and 7 apples cost Katie \$4.90.
Calculate the prices of peaches and apples.
Call the price of peaches **p**.
Call the price of apples **a**.

$6p + 5a = 5.70$ ①
$4p + 7a = 4.90$ ②
Multiply ① by -7: $-42p - 35a = -39.90$ ③
Multiply ② by 5: $20p + 35a = 24.50$ ④
Add ③ and ④: $-22p = -15.4$
$\mathbf{p = 0.70}$
Substitute for p in ①: $6(0.70) + 5a = 5.70$
$5a = 5.70 - 4.20$
$5a = 1.50$
$\mathbf{a = 0.30}$

∴ **Peaches cost \$0.70 and apples cost \$0.30.**

Write and solve simultaneous equations to find the unknown numbers.

1 Track pants cost \$16 less than 3 pairs of shorts. Two pairs of shorts and 3 pairs of track pants cost \$150. Calculate the prices of shorts and track pants.

ISBN: 9780170438568

2 An ice cream costs 5 times the price of a juicy. Four ice creams and 5 juicies cost \$17.50. Calculate the prices of ice creams and juicies.

3 There are twice as many boys as girls in Erin's class of 27 students. How many boys and girls are there in her class?

4 The number of Merit grades that Kiri has is 5 more than double the number of her Achieved grades. If she doubled her Merit grades and added 4, she would have 6 times the number of her Achieved grades. How many of each grade does she hold at present?

5 Bert has 4 times as many marbles as Sadie. If Sadie had 42 more, she would have twice as many as Bert has. How many marbles does each of them have?

6 Kristen bought 5 lollipops and 2 chocolate bars for \$3.30. Hugh bought 3 lollipops and 4 chocolate bars for \$4.50. Calculate the prices of lollipops and chocolate bars.

ISBN: 9780170438568

7 A pohutukawa tree costs 3 times the price of a tree fern. Seven pohutukawa trees and 12 tree ferns cost $231. Calculate the prices of pohutukawa trees and tree ferns.

8 Six peaches and 5 apples cost $5.70. Four peaches and 7 apples cost $4.90. Calculate the prices of peaches and apples.

9 Three pies cost the same as 2 burgers. Five burgers and 4 pies cost $27.60. Calculate the prices of burgers and pies.

10 Hoodies cost $10 more than 2 T-shirts. Five hoodies and 11 T-shirts cost $449. Calculate the prices of hoodies and T-shirts.

11 Maria sold boxes of chocolate bars to help raise funds for a sports trip. Each box earned her $7. Matiu sold boxes of soap, which earned him $4.50 per box. Altogether they earned $127.50. If they sold a total of 25 boxes between them, how many boxes did each student sell?

ISBN: 9780170438568

Rearrangement of expressions

1 Where the subject appears once

- First put **all** terms that contain the required subject on the left.
- If the subject appears in several terms, collect all of these to one side and factorise.
- Get rid of fractions by multiplying by the denominator.
- Get rid of square roots, isolating the square root on one side and then squaring both sides.
- Use normal equation-solving rules to isolate the subject.

Examples:

1 Make r the subject of $V = \pi r^2 h$.

$\pi r^2 h = V$ **(÷ by πh)**

$r^2 = \frac{V}{\pi h}$ **(√ both sides)**

$r = \sqrt{\frac{V}{\pi h}}$

2 Make w the subject of $A = \pi\sqrt{\frac{w}{g}}$.

$\pi\sqrt{\frac{w}{g}} = A$ **(÷ by π)**

$\sqrt{\frac{w}{g}} = \frac{A}{\pi}$ **(square)**

Get the √ on its own on the left.

$\frac{w}{g} = \frac{A^2}{\pi^2}$ **(x by g)**

$w = \frac{A^2 g}{\pi^2}$

3 Rewrite the formula $A = \frac{(a+b)}{2}h$ with a as the subject.

$\frac{(a+b)}{2}h = A$ **(x by 2)**

$(a+b)h = 2A$ **(expand brackets)**

$ah + bh = 2A$ **(– bh)**

$ah = 2A - bh$ **(÷ by h)**

$a = \frac{2A - bh}{h}$

4 Express a in terms of b if $(a^4)^3 = (b^3)^2$.

$(a^4)^3 = (b^3)^2$

$a^{12} = b^6$

$a^{12} = (b^{\frac{1}{2}})^{12}$

$a = b^{\frac{1}{2}}$ or $\sqrt{b}$

Rearrange the following equations.

1 Make r the subject of $A = \pi r^2$.

____

____

____

____

2 Make P the subject of $I = \frac{PRT}{100}$.

____

____

____

____

3 Rearrange $y = mx + c$ so that x is the subject.

____

____

____

____

4 The formula for velocity is $v = \frac{d}{t}$. Make t the subject of this formula.

____

____

____

____

ISBN: 9780170438568

5 Rewrite the formula $A = \frac{(a + b)}{2}h$ with h as the subject.

6 Rewrite the expression $y = ax^2 + b$ with x as the subject.

7 The sums of the areas of two triangles that have the same height is given by $A = \frac{1}{2}h(B + b)$. Write this formula with h as the subject.

8 The sums of the areas of two triangles that have the same height is given by $A = \frac{1}{2}h(B + b)$. Write this formula with B as the subject.

9 Make h the subject of $V = \frac{1}{3}r^2h$.

10 Make r the subject of $V = \frac{1}{3}r^2h$.

11 The total surface area of a cylinder is $A = 2\pi r(r + h)$. Make h the subject of this formula.

12 The formula for converting degrees Fahrenheit into degrees Celsius is $F = 32 + \frac{9C}{5}$. Make C the subject.

13 Express a in terms of b if $a^2 \times a^4 = (b^3)^2$.

14 Express a in terms of b if $(a^2)^3 \times a^4 = b(b^2)^2$.

15 Express a in terms of b if $\frac{(a^3)^2}{a^2} = (b^4)^2$.

16 Express a in terms of b if $(a^{10})^{\frac{1}{2}} = (b^{\frac{1}{2}})^{20}$.

ISBN: 9780170438568

2 Where the subject appears twice

- If the subject appears in several terms, collect all of these to one side and factorise.
- Get rid of fractions by multiplying by the lowest common multiple (LCM) of the denominators.

Examples:

1 Make a the subject of $5ab - 3 = 2a$.

$5ab - 3 = 2a$ **(– 2*a*)**

$5ab - 2a = 3$ **(factorise)**

$a(5b - 2) = 3$ **(÷ by (5*b* – 2))**

$a = \frac{3}{5b - 2}$

Both terms with *a* in them on the left.

2 Make a the subject of $\frac{1}{b} = \frac{2}{a} + \frac{3}{c}$.

$\frac{1}{b} = \frac{2}{a} + \frac{3}{c}$ **(x by $\frac{abc}{1}$)**

$\frac{1}{b} \times \frac{abc}{1} = \frac{2}{a} \times \frac{abc}{1} + \frac{3}{c} \times \frac{abc}{1}$

$ac = 2bc + 3ab$ **(– 3*ab*)**

$ac - 3ab = 2bc$ **(factorise)**

$a(c - 3b) = 2bc$ **(÷ by (*c* – 3*b*))**

$a = \frac{2bc}{c - 3b}$

Make a the subject of each of the following expressions.

1 $2a = 6 - ab$

2 $5a = 3b - ab$

3 $2(a - 1) = ab$

4 $2(3a - b) = 5b(1 - a)$

5 $b + 1 = \frac{4}{a}$

6 $\frac{3}{a} = 2 + \frac{1}{b}$

7 $\frac{1}{a} - \frac{1}{b} = \frac{1}{c}$

8 $\frac{1}{3} = \frac{2}{a} - \frac{1}{b}$

ISBN: 9780170438568

3 Problems involving the rearrangement of expressions

It is often easiest to begin with an **equation written with words**.

Examples:

1 The volume of a cone is given by $V = \frac{1}{3}r^2h$. A second cone is created with the same height but twice the volume. Call the radius of this second cone R.
Write an expression for the radius of the larger cone R in terms of r.

vol big cone = 2 x vol small cone

$\frac{1}{3}\pi R^2h = 2(\frac{1}{3}\pi r^2h)$ **(x by 3)**

$\pi R^2h = 2\pi r^2h$ **(÷ by πh)**

$R^2 = 2r^2$ **(take √)**

$R = \sqrt{2r^2}$

or $\sqrt{2}\,r$

2 The vertical height of this triangle is given by $a = \sqrt{h^2 - 10^2}$.
A second larger triangle is constructed with the same base (10), but it is three times as high.
Find an expression for the hypotenuse of the new triangle (H) in terms of h.

Height of taller Δ = 3 x height of shorter Δ

$\sqrt{H^2 - 10^2} = 3\sqrt{h^2 - 10^2}$ **(square)**

$H^2 - 10^2 = 9(h^2 - 10^2)$ **(expand, + 100)**

$H^2 = 9h^2 - 800$ **(take √)**

$H = \sqrt{9h^2 - 800}$

1 The volume of a cone is given by $V = \frac{1}{3}\pi r^2h$. A second cone is created with the same radius but three times the volume. Call the height of this second cone H.
Write an expression for the height of the larger cone H in terms of h.

____

____

____

2 The length of the hypotenuse of this triangle is given by $l = \sqrt{b^2 + 36}$.
A second larger triangle is constructed with the same height (6) but its hypotenuse is three times as long.
Find an expression for the base of the new triangle (B) in terms of b.

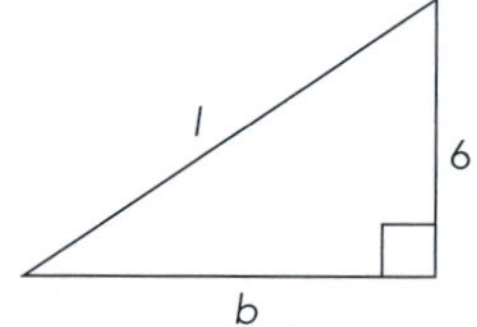

____

____

____

3 The length of the base of this triangle is given by $b = \sqrt{l^2 - 36}$. A second larger triangle is constructed with the same height (6) but its base is twice as long. Find an expression for the length of the hypotenuse of the larger triangle (L) in terms of l.

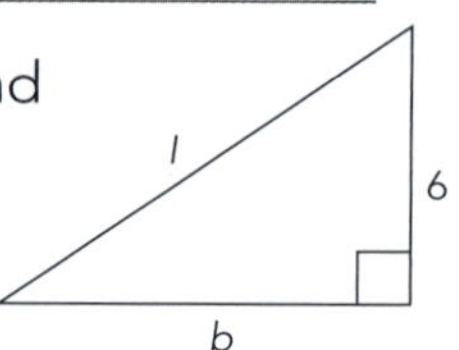

____

____

____

ISBN: 9780170438568

4 A cylinder and a sphere have the same volume and the same radius.
The volume of a cylinder is $V = \pi r^2 h$ and the volume of a sphere is $V = \frac{4}{3}\pi r^3$.
Describe the relationship between the height of the cylinder and its radius.

5 A cylinder and a sphere have the same surface area and the same radius.
The surface area of a cylinder is $SA = 2\pi r^2 + 2\pi rh$ and the surface area of a sphere is $SA = 4\pi r^2$. Describe the relationship between the height of the cylinder and its radius

6 The curved surface of a cone and the curved surface of a hemisphere have the same surface area and the same radius.
The curved surface area of the cone is $SA = \pi r\sqrt{h^2 + r^2}$ and the curved surface area of a hemisphere is $SA = 2\pi r^2$. Describe the relationship between the height of the cone and its radius.

7 A circle fits inside an isosceles trapezium so that its height is $2r$. The longer parallel side of the trapezium is twice the length ($2x$) of the shorter side (x). The area of a trapezium is given by the formula $A = \frac{a+b}{2}h$, and the area of a circle is $A = \pi r^2$. If the shaded area is the same as the area of the circle, find the relationship between the shorter parallel side (x) and the radius of the circle (r).

ISBN: 9780170438568

Algebraic proof

- An algebraic proof is a mathematical argument that usually does not use numbers.
- Often the question will ask you to 'Show that …'

Hints:

- Even numbers: call these $2x$, e.g. 10, where $x = 5$.
- Odd numbers: call these $2x + 1$, e.g. 15, where $x = 7$.
- Consecutive numbers come after each other: call these x, $x + 1$, $x + 2$, e.g. 6, 7, 8 where $x = 6$.
- Consecutive even numbers: call these $2x$, $2x + 2$, $2x + 4$, e.g. 6, 8, 10, where $x = 3$.
- Consecutive odd numbers: call these $2x + 1$, $2x + 3$, $2x + 5$, e.g. 9, 11, 13, where $x = 4$.
- A prime number can only be divided by two numbers: 1 and itself. Therefore 1 is not a prime number.

Examples:

1 Show that the difference between two odd numbers is always even.

Let the two odd numbers be $2x + 1$ and $2y + 1$

$$\begin{aligned}\text{Difference} &= (2x + 1) - (2y + 1)\\ &= 2x + 1 - 2y - 1\\ &= 2x - 2y\\ &= 2(x - y)\end{aligned}$$

Any number multiplied by 2 is even.
$\therefore$ The difference between two odd numbers must be even.

2 Show that for any number n, $n^2 + n$ can never be prime.

Let n be even, and let $n = 2x$:

$$\begin{aligned}(2x)^2 + (2x) &= 4x^2 + 2x\\ &= 2(2x^2 + x)\end{aligned}$$

As a multiple of 2, this cannot be prime.

Let n be odd, and let $n = 2x + 1$:

$$\begin{aligned}(2x + 1)^2 + (2x + 1) &= 4x^2 + 4x + 1 + 2x + 1\\ &= 4x^2 + 6x + 2\\ &= 2(2x^2 + 3x + 1)\end{aligned}$$

As a multiple of 2, this cannot be prime.

$\therefore$ $n^2 + n$ can never be prime.

1 Show that the sum of three consecutive whole numbers is always divisible by 3.

____

____

____

____

____

2 Show that the product of two even numbers is always a multiple of 4.

____

____

____

____

____

 ISBN: 9780170438568

3 Show that the product of two odd numbers is always odd.

4 If you find the product of two consecutive whole numbers and then add the larger number, the answer is always a square number. Use algebra to show that this is true.

5 Show that if n is a whole number, $n^2 + n + 2$ must always be even.

6 Show that for any two-digit whole number (ab) that is less than 100, if the digits add to 9, then the number (ab) must be a multiple of 9.

7 Use algebra to show that any whole number under 100 that ends with 5 has a square that ends with 25.

ISBN: 9780170438568

General pattern and formula problems

1 Rectangular patterns are created with dots as shown below.

Pattern number (n)	1	2	3	4	5
Number of dots (D)	2	6	12	20	30

The formula for the total number of dots (D) is given by: $D = n(n + 1)$

a How many dots would be needed for the 20th pattern?

b Which pattern number would have 110 dots? Explain your answer fully.

2 Paint covers about 11 m^2 per litre. The formula for calculating the total cost (T) of painting a rectangular surface is $T = \frac{CLWP}{11}$, where:

C represents number of coats required
L represents the length of the surface
W represents the width of the surface
P represents the price of the paint per litre

a Calculate the total cost of doing two coats of paint that costs $33 per litre, on a wall that is 10 m long and 2 m high.

b It costs $90 to paint a wall that is 5.5 m long and 2 m high, with paint that costs $30 per litre. How many coats of paint were needed?

c Find an expression for the number of coats applied (C) in terms of the total cost, the paint price, and the width and height of the surface being painted.

ISBN: 9780170438568

3 The mean (M) of three numbers (a, b and c) is given by the formula $M = \frac{a+b+c}{3}$.

a Calculate the mean of 17, 8 and 11.

b The mean of 5, 19 and a third number is 15. Calculate the third number.

c Write an expression for finding c in terms of the mean and the other two numbers.

4 Kyle sells sports equipment. He is paid \$300 per week. In addition he is paid commission, which depends on how much he sells. He gets no commission on the first \$2000 worth each week. He gets 10% commission on the value of everything he sells after he has sold equipment worth \$2000.

a Write an expression for T, the total amount he gets paid, in terms of x, the total amount he sells.

b Calculate his total pay if he sells \$3600 worth of equipment in a week.

c In a really good week he was paid \$680. Calculate the total value of the equipment he sold.

d Write an expression for the value of equipment sold (x) in term of the total (T) that Kyle earned in a week.

ISBN: 9780170438568

5 Spiral patterns are created from matchsticks as shown.

Pattern number (n)	1	2	3	4	5
Number of matchsticks (M)	2	4	7	11	16

The formula for the total number of matchsticks (M) is given by: $M = 0.5n^2 + 0.5n + 1$.

a How many matchsticks will be used in the 20th pattern?

b Which pattern would have 46 matchsticks? Explain your answer fully.

c If $M = 0.5n^2 + 0.5n + 1$ and $T = (3n + 2)(n - 1) - (n - 1)^2 + (n + 7)$, write an equation for T in terms of M.

6 A pattern of numbers is shown in the table, but two of the numbers are missing.

n	1	2	3	4
y	1			64

Suggest what the pattern could be: for each suggestion, write an equation and list the missing numbers.

 ISBN: 9780170438568

7 The surface area (S) of a cuboid is given by the formula $S = 2lw + 2wh + 2lh$, where

w represents the width of the cuboid
l represents the length of the cuboid
h represents the height of the cuboid

a Calculate the surface area of a cuboid that has a width of 4 cm, a length of 5 cm and a height of 2 cm.

b Another cuboid has a surface area of 104 cm^2. It is 2 cm high and 6 cm long. Calculate its width.

c Find an expression for the length of a cuboid in terms of its width, height and surface area.

8 These patterns are created from matchsticks as shown.

Pattern number (n)	1	2	3	4	5
Number of matchsticks (M)	6	12	20	30	42

The formula for the total number of matchsticks (M) is given by: $M = n^2 + 3n + 2$.

a Which pattern needs 90 matches? Explain your answer fully.

b If $M = n^2 + 3n + 2$ and $T = (2n + 1)^2 + (n + 4)^2 - (7 - 3n)$, write an equation for T in terms of M.

ISBN: 9780170438568

Algebra with rectangles

All of the following figures are based on rectangles.

1

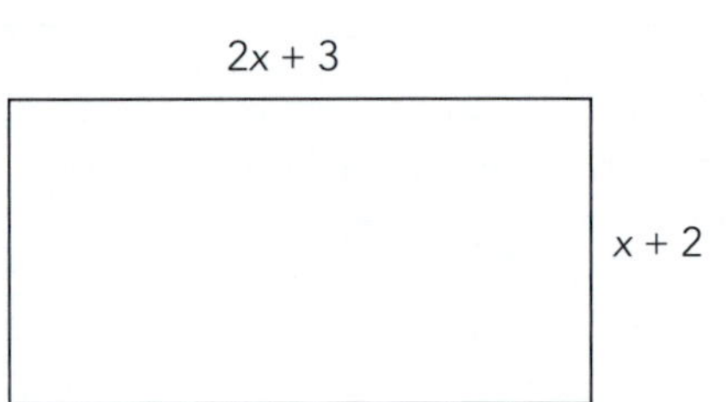

a Write an expression for the perimeter of the rectangle.

b If the perimeter is 28 cm, find the dimensions of the rectangle.

c Write an expression for the area of the rectangle.

2

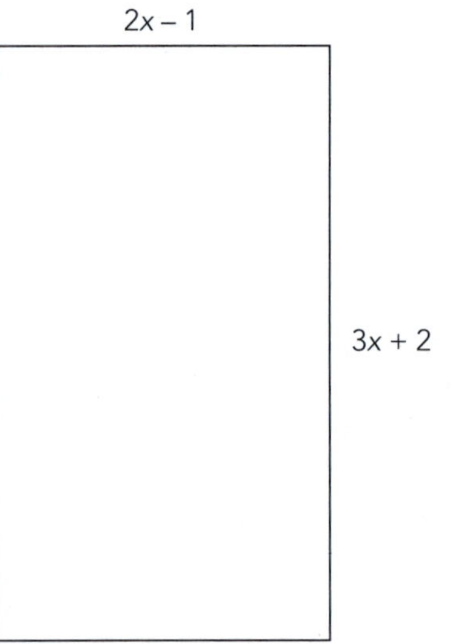

a Write an expression for the perimeter of the rectangle.

b If the perimeter is 52 cm, find the dimensions of the rectangle.

c Write an expression for the area of the rectangle.

3

a The area of the rectangle is $x^2 + 2x - 3$. Find an expression for the longer side.

b If the area of the rectangle is 96 cm^2, find its dimensions.

 ISBN: 9780170438568

4

a The area of the rectangle is $x^2 + x - 20$. Find an expression for the longer side.

____

____

b If the area of the rectangle is 36 cm^2, find its dimensions.

____

____

5

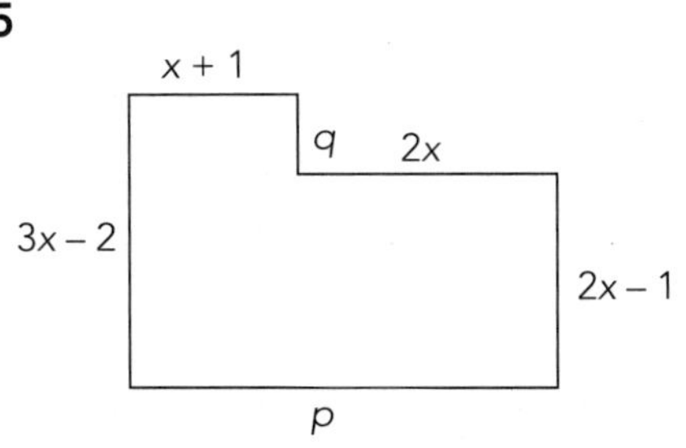

a Write lengths p and q in terms of x.

p = _______________

q = _______________

b Write an expression for the perimeter of the figure.

____

____

c If the perimeter is 34 cm, find the value of x.

____

____

d Write an expression for the area of the figure.

____

____

6

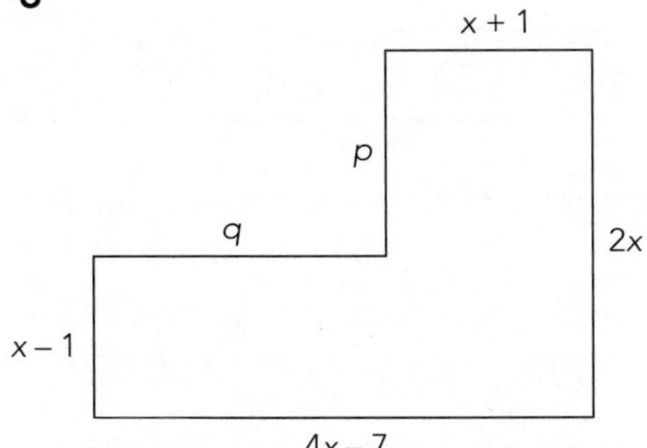

a Write lengths p and q in terms of x.

p = _______________

q = _______________

b Write an expression for the perimeter of the figure.

____

____

c If the perimeter is 46 cm, find the value of x.

____

____

d Write an expression for the area of the figure.

____

____

ISBN: 9780170438568

Practice tasks

Practice task 1

a **i** The product of two numbers is $2x^2 - 3x - 20$. Write expressions for each of the two numbers in terms of x.

ii If the product of the two numbers is positive, what would be possible values of x?

b **i** Hemi knows that the formula for finding the volume (V) of a cylinder is given by the formula $V = \pi r^2 h$, where r is the radius and h is the height of the cylinder. Write a formula he could use to find the radius of a cylinder in terms of its volume, height and π.

ii Use your formula to find the radius of a cylinder that is 6 cm high and has a volume of 54π cm^3.

  ISBN: 9780170438568

c A cuboid has a volume equal to $4x^2 - 10x - 6$.

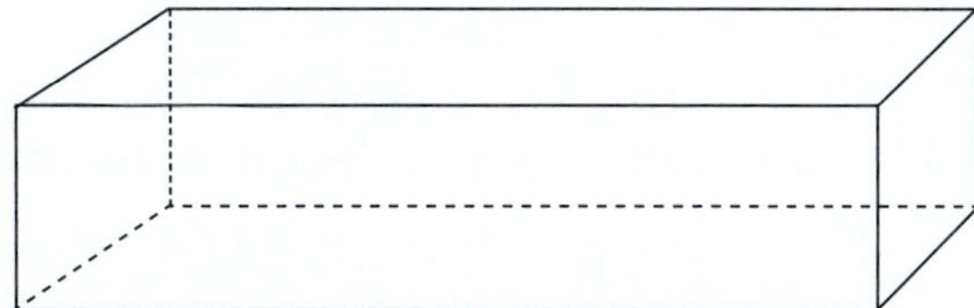

i Find the dimensions of the cuboid in terms of x.

ii If x can be only whole numbers, what is the smallest volume possible of the cuboid?

d Pita has found a magic trick. He thinks of a two-digit number. He adds the digits together, and then subtracts the result from the original number. He has discovered that the answer is always a multiple of nine. Use algebra to show that the trick always works.

ISBN: 9780170438568

Practice task 2

a Two sides of a triangle are given by $3(2x + 1)$ and $7(x - 2)$.
If its perimeter is $x^2 + 12x - 13$, write down the length of the third side.

b A rectangle has one side with a length $5(2 + 3y)$. If its area is given by the expression $40x^3y + 60x^3y^2$, write down the length of the other side.

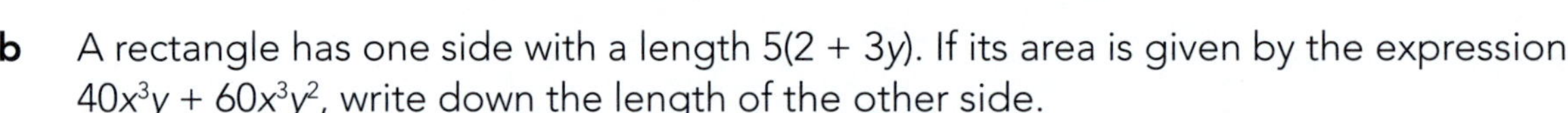

c Solve 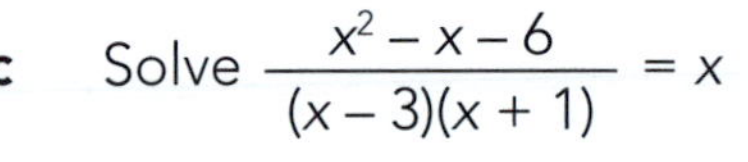 $\dfrac{x^2 - x - 6}{(x - 3)(x + 1)} = x$

d Lavini thinks of a number. She squares it, and then adds seven times the original number. The result is 60. What numbers could she have been thinking of?

ISBN: 9780170438568

e $A = 2x(x - 1) + x(x + 5)$ and $B = (3x - 2)(x + 1) + 2(1 - x^2)$.
Write an expression for A in terms of B.

f Three adult tickets to a movie can be bought for the price of five children's tickets. If an adult ticket was $1.70 more, it would cost double a child's ticket.
Find the prices of adults' and children's tickets.

g Consider the following pair of equations:

$$3x - 2y = 5$$
$$6x + py = 12$$

Give a value for p for which there would be no solutions to this pair of equations.
Explain your reasoning carefully.

ISBN: 9780170438568

Practice task 3

a Pepa is making a pattern with groups of buttons. She begins with one button, her second group has three buttons, her third group has six buttons, and so on. She wants to make only whole groups with no buttons left over.

Pattern number (n)	**1**	**2**	**3**	**4**	**5**
Triangular number (T)	1	3	6	10	15

The formula for this sequence is $T = \frac{1}{2}n(n + 1)$.

i How many buttons will she need for the tenth group?

ii If she has 45 buttons, how many groups can she make?

iii Explain why $B = 50$ does not give solutions for n in this context.

ISBN: 9780170438568

b Show that $\frac{2(3p + 2)}{6} - \frac{1}{p}$ is the same as $\frac{3p^2 + 2p - 3}{3p}$.

c If *m* is a whole number, for what values of *m* is $2^{(2m-3)} > 70$?

d Every weekend, Tara is paid \$45 to referee up to four football games. If she has to referee more than four games, she is paid \$14 per game.

Let g = the number of games she referees.

Write a formula so that she can work out the number of games she has refereed from the amount she was paid (P).

e Make x the subject of $2 - \frac{5}{x} = 7y$.

ISBN: 9780170438568

Answers

Simplifying expressions (pp. 7–8)

Multiplying and dividing (p. 7)

1 y^3 **2** adf **3** $7d$
4 $6df$ **5** $-3efg$ **6** $7df^2$
7 $-10a^2$ **8** $-24abc$ **9** $-5cd^2$
10 $18f$ **11** $\frac{g}{2}$ **12** $\frac{14d}{e}$
13 $\frac{3}{e}$ **14** $\frac{20dg}{f}$ **15** $\frac{12df}{ac}$
16 $\frac{12efg}{a^3}$

Adding and subtracting (p. 8)

1 like **2** unlike
3 unlike **4** like
5 like **6** like
7 like **8** unlike
9 unlike **10** like
11 $4x$ **12** $6x^2 + x$
13 $5 + 2ab$ **14** $9p^5q + 3pq^5$
15 $8y^2 - y^3$ **16** $x^2 + 12x + 35$
17 $11x - 4y + 1$ **18** $6ab + 4a - 3b$
19 $17 + q - 6p$ **20** $x^2 - 12x + 27$
21 $y^2 + 7y - 8$ **22** $3x^2 + 6xy - 7y^2$
23 x^2 **24** $5x^2yz + xyz^2$

Exam questions (p. 9)

1 $y^2 + 4y - 3$
2 $2x^2 - 3z + 1$
3 $a = 5x - 3$
$b = 4x^2 - 9$
Perimeter $= 8x^2 + 14x - 24$
4 Fifth side $= x^2 + 4xy - 6y^2 - 5x$

Powers (pp. 10–16)

Multiplying terms that have powers (p. 10)

1 p^5 **2** a^5 **3** b^9
4 g^9 **5** $6a^3$ **6** $10a^3b^2$
7 $30b^{15}$ **8** f^8g^7 **9** $15y^{10}$
10 $-a^3b^4$ **11** $-6y^6$ **12** $8z^{10}$

Dividing terms that have powers (p. 11)

1 y^6 **2** y^3 **3** b^9
4 $5b^5$ **5** $\frac{y^3}{2}$ **6** x^2y
7 $4y^2z$ **8** $-\frac{1}{2x}$ **9** $\frac{3x}{y^5}$
10 $-\frac{x}{3}$ **11** $\frac{1}{3x}$ **12** $\frac{2y^2z^2}{x^3}$
13 $\frac{1}{6}$ **14** $\frac{1}{2xz}$

Powers of powers (p. 12)

1 b^8 **2** $16x^{12}$ **3** a^{14}
4 $-y^{15}$ **5** $9x^8$ **6** $-64x^{12}$
7 $-50x^6$ **8** $x^6y^{12}z^9$ **9** $-9a^6b^4$
10 $18x^2y^6$ **11** $\frac{x^6}{8}$ **12** $\frac{x^8}{25}$

Mixing it up (p. 13)

1 b^4 **2** 1 **3** c^2b^3
4 $\frac{9}{b}$ **5** $\frac{a^6}{8}$ **6** $\frac{16z^4}{25}$
7 $64z^{12}$ **8** $\frac{1}{25y^6}$ **9** $54y^7z^{13}$
10 $\frac{64z^{14}}{y}$

Roots (p. 14)

1 5 **2** x^5 **3** 10
4 $6x^3$ **5** $4x^8$ **6** 4
7 $10x^{50}$ **8** $7x^4y^3$ **9** $9x^{64}$
10 $8x^6y^{12}$ **11** $5x^5y^3$ **12** $8x^{32}y^8z^{18}$

Exam questions (pp. 15–16)

1 Area $= 12a^3b^4c$ **2** Area $= 9p^8q^2$
3 Volume $= 6x^7y^5z$ **4** Length $= 4x^8y^2$
5 Length $= 4x^6y^3$ **6** Volume $= 125x^6y^3$
7 Length $= 6x^2y^7z$ **8** Length $= 4a^2b^4$
9 Length $= 0.5a^8b^{18}$ **10** Volume $= \pi f^5g^5$

Expanding and factorising (pp. 17–20)

Expanding brackets (p. 17)

1 $15x - 20$ **2** $12x - 6x^2$
3 $-6x^2 + 10x$ **4** $21x - 16$
5 $6x - 12$ **6** $16x - 21$
7 $2 - 6x$ **8** $24x + 23$
9 $2x^2 - 8x + 4$ **10** $23x - 24$
11 $14x^2 - 6x$ **12** $-23x^2 - 27x$

Factorising (pp. 18–19)

1 $8(2x + 1)$ **2** $12(3x - 5)$
3 $20(5 + x)$ **4** $12x(5 - 2x)$
5 $4xy(3y + 1)$ **6** $ab^3c^3(b + ac)$
7 $5(x - 3)$ **8** $x(x - 3)$
9 $8(x + 2)$ **10** $6(4 - 3a)$

ISBN: 9780170438568

11 $2(5y + 3)$
12 $x(x + 11)$
13 $x(2 + 3x)$
14 $5x(x + 3)$
15 $4y(y + 3)$
16 $2a(2a + 1)$
17 $3x(x + 5)$
18 $7y(3 - 2y^2)$
19 $2a^2(5a + 1)$
20 $4b^2(3 + 4b)$
21 $xy(1 + 9xy)$
22 $7xy(2x + y)$
23 $abc(2b - ac^2)$
24 $8abc^3(3b - a)$
25 $x(10y + 5x - y^2)$
26 $4xy(3y + x - 2xy)$

Exam questions (p. 20)

1 Perimeter = $42p - 12$ or $6(7p - 2)$
2 Perimeter = $22t + 18$ or $2(11t + 9)$
3 Length = $4x + 2$ or $2(2x + 1)$
4 Length = $3 - 5a$
5 Length = $3x^2y$
6 Mean = $3x^2 + 2x - 4$
7 Mean = $y^2 + 2xy - 3y$

Algebraic fractions (pp. 21–26)

Multiplying fractions (p. 21)

1 $\frac{2y}{9}$ 2 $\frac{14}{z^2}$ 3 $\frac{8z}{35}$
4 $\frac{18x}{y^2}$ 5 $\frac{5}{6z}$ 6 $\frac{y}{6}$
7 $\frac{3(y-5)}{20}$ 8 $\frac{y^2z^3}{7}$ 9 $\frac{4z}{15}$
10 $\frac{3x(x+1)}{10}$ 11 $\frac{2xy^2z^2}{5}$ 12 $\frac{y}{2(y-1)}$

Dividing fractions (p. 22)

1 6 2 $\frac{y}{x}$ 3 $\frac{3}{2}$
4 $3y$ 5 25 6 $12x^2$
7 $\frac{1}{y^2}$ 8 $\frac{5}{2}$ 9 $\frac{12x}{5}$
10 $\frac{25x}{y}$

Adding and subtracting fractions (pp. 23–24)

1 Where the denominators are the same (p. 23)

1 $\frac{5x}{9}$ 2 $\frac{9}{x}$ 3 $\frac{x}{7}$
4 $\frac{6x}{y}$ 5 $\frac{5x}{11}$ 6 $\frac{x}{2}$
7 x 8 $\frac{6}{x}$ 9 $\frac{6x-1}{7}$
10 $\frac{11x-1}{9}$ 11 $\frac{x-2}{2}$ 12 $1 - x$

2 Where the denominators are different (p. 24)

1 $\frac{11x}{12}$ 2 $\frac{13x}{14}$ 3 $\frac{7x}{15}$
4 $\frac{x}{14}$ 5 $\frac{5xy-3}{15}$ 6 $\frac{24-x}{6}$
7 $\frac{3y-2x}{12}$ 8 $\frac{13+x}{5}$ 9 $\frac{7x-5}{5}$
10 $\frac{4x-5}{6}$

3 Where the denominators are different algebraic equations (p. 25)

1 $\frac{7x-5}{(x+1)(2x-4)}$ 2 $\frac{-3x^2+5x+3}{(3x-2)(x+1)}$
3 $\frac{2x^2+8x+14}{(x+1)(x+5)}$ 4 $\frac{7x^2-8x+10}{(3x-2)(x+4)}$

Exam questions (p. 26)

1 $a = 3$, $b = xyz$
2 $\frac{8p}{5q} \div \frac{p^2}{10} = \frac{8p}{5q} \times \frac{10}{p^2}$
$= \frac{16}{pq}$
∴ he is not correct.
3 $a = 5$, $b = 2pq^3$
4 $a = 3y - 5$, $b = x$
5 Area = $\frac{35x^2 + 6x}{15}$
6 $\frac{2}{x} + \frac{3(1-x)}{y} = \frac{2y}{xy} + \frac{3x(1-x)}{xy}$
$= \frac{2y + 3x - 3x^2}{xy}$
∴ she is correct.
7 $a = x^2 + 10x - 16$

Substitution (p. 27)

1 -3 2 9 3 48
4 -5 5 -2 6 -1
7 31 8 2 9 37
10 36π 11 36π 12 $27 - 9\pi$

Solving linear equations (pp. 28–33)

One-step equations (p. 28)

1 1.7 2 3.1 3 2.8
4 2 5 -47 6 -0.68
7 11.5 8 0.6 9 0.1
10 -133

Two-step equations (p. 29)

1 3.8 2 10 3 3
4 41.75 5 -20 6 -0.04
7 0.2 8 16 9 47
10 0.005

Equations with a variable on both sides (p. 30)

1 10 2 -2 3 9
4 -5 5 20 6 3
7 5 8 -122 9 4
10 -0.1

Equations with brackets (p. 31)

1 -22 2 -3 3 -1
4 1.5 5 1 6 1
7 3 8 2.2 9 1
10 0.5

Equations with fractions (p. 32)

1 35 2 10 3 3.75
4 5 5 -0.6 6 0.25
7 1 8 7

Forming and solving linear equations (p. 33)

1 $15a + 7(9 - a) = 87$
$a = 3$, so there were 3 adults and 6 children
2 $6s + 9(s + 1.5) = 43.5$
$s = 2$, so small cost \$2.00, large cost \$3.50
3 $3H = H + 22$
$H = 11$, so Hine will be 11, her mother will be 33
4 $44 - x = 3(20 - x)$
$x = 8$, so Dan was 12 and his father 36
5 $2(x + 3) + x = 39$
$x = 11$, so dimensions are 14 cm, 14 cm and 11 cm
6 $4x - 14 = 62$
$x = 19$, so dimensions are 19 cm, 12 cm, 19 cm and 12 cm
7 $2x + x + 31 = 118$
$x = 29$, so max dimensions are 29 cm by 58 cm by 31 cm
8 $19 - r = 1.5(r + 5) - 1$
$r = 5$, so she gave him 5 cherries

Inequations (pp. 36–37)

1 $x > 8$ 2 $x < 11$
3 $x \geq 4$ 4 $x < -7$
5 $x > 1$ 6 $x \geq -43$
7 $x > -5.5$ 8 $x \leq 1$

Forming and solving inequations (p. 37)

1 $11 - 4x > 1$
$x < 2.5$
2 $x + 1.5x + x + 1.5x \leq 75$
Maximum dimensions are 22.5 cm by 15 cm
3 $m + m + (m - 3) \leq 60$
$m = 21$, so Mick's and Adam's meals cost \$21 each, Sam's cost \$18
4 $100 - 8.5x \geq 35$
$x \leq 7.64$, so she can invite 6 friends (don't forget she needs a ticket too!)
5 $p \geq \$4.50 + \frac{2}{5}s$
$p = 9.10$, so Penny has \$9.10

Equations with powers (p. 38)

1 9 2 3 3 2 4 -4
5 2 6 7 7 -3 8 2
9 6 10 -4

Quadratic expressions (pp. 39–61)

Expanding quadratic expressions (pp. 39–40)

1 $x^2 + 7x + 12$ 2 $x^2 + 5x - 14$
3 $x^2 - 6x - 27$ 4 $x^2 - 13x + 30$
5 $x^2 + 13x + 40$ 6 $3x^2 - 20x - 7$
7 $x^2 + 16x + 64$ 8 $x^2 - 10x + 25$
9 $3x^2 - 5x - 2$ 10 $6x^2 + 16x - 6$
11 $9 - 6x + x^2$ 12 $9x^2 - 30x + 25$
13 $6x^2 - 15x$ 14 $x^2 - 1$
15 $20x^2 + 7x - 6$ 16 $28 + 13x - 6x^2$
17 $16x^2 - 9$ 18 $25 - 4x^2$

Factorising quadratic expressions (pp. 41–45)

1 Where the coefficient of x^2 is 1 (pp. 41–42)

1 $(x + 2)(x + 4)$ 2 $(x + 7)(x + 1)$
3 $(x + 4)(x + 9)$ 4 $(x + 6)(x - 2)$
5 $(x + 15)(x + 4)$ 6 $(x - 6)(x - 3)$
7 $(x + 4)(x - 6)$ 8 $(x - 5)^2$
9 $(x + 1)(x - 6)$ 10 $(x + 2)(x - 3)$
11 $(x - 2)(x - 3)$ 12 $(x + 6)(x - 1)$
13 $(x + 5)(x - 5)$ 14 $(7 + x)(7 - x)$
15 $(3 + 4x)(3 - 4x)$ 16 $(5 + 3x)(5 - 3x)$
17 $(3 + x)(8 + x)$ 18 $(3 - x)(5 - x)$

2 Where the coefficient of x^2 is not 1, but there is a common factor (p. 43)

1 $4(x + 2)(x + 6)$ 2 $2(x + 3)(x + 4)$
3 $5(x + 2)(x - 3)$ 4 $3(x + 7)(x - 2)$
5 $10(x - 2)^2$ 6 $5(x + 6)(x - 6)$
7 $5(x - 1)^2$ 8 $8(x + 1)(x - 1)$
9 $2(x + 1)(x - 3)$ 10 $8(5 + x)(5 - x)$

3 Where the coefficient of x^2 is not 1, and there is no common factor (pp. 44–45)

1 $(2x + 3)(x + 4)$ 2 $(2x + 1)(x + 6)$
3 $(2x + 3)(2x + 5)$ 4 $(2x + 3)(x - 4)$
5 $(3x - 2)(x - 6)$ 6 $(2x - 3)(x - 1)$
7 $(x - 5)(4x + 5)$ 8 $(3x + 1)(2x - 3)$
9 $(2x + 3)(5x - 1)$ 10 $(3x - 2)(3x - 1)$
11 $(3x + 5)(2x + 7)$ 12 $(7x - 2)(3x - 4)$

Simplifying quadratic fractions (p. 46)

1 2 2 $\frac{1}{4}$ 3 x
4 $\frac{2}{3}$ 5 $x + 8$ 6 $x - 4$
7 $\frac{1}{x - 3}$ 8 $\frac{3}{x + 5}$ 9 $\frac{x + 1}{x - 1}$
10 $\frac{x + 2}{x + 4}$

ISBN: 9780170438568

Exam questions (p. 47)

1 Area = $12x^2 + 32x - 35$
2 a = 15, b = –22, c = 8
3 Length = $7x + 2$, Width = $2x - 1$
4 $a = 5x - 2$ and $b = x + 4$ (or the other way around)
5 Length = $4x + 5$, Width = $4x - 5$
6 Length = $5x - 2$
7 Height = $7x - 3$
8 a = 2, b = –3, c = 4, d = 5 or a = 4, b = 5, c = 2, d = –3

Solving quadratic equations (pp. 48–52)

1 Factorised quadratic equations (pp. 48–49)

1 $x = -9$ or 1
2 $x = 7$ or -2
3 $x = -1$ or -4
4 $x = 8$ or 3
5 $x = -7$ or 3
6 $x = 3$
7 $x = -6$
8 $x = -\frac{1}{2}$
9 $x = -2$ or 5
10 $x = -4$ or 3
11 $x = 0$ or 1
12 $x = 0$ or -9
13 $x = 0$ or $\frac{1}{3}$
14 $x = 0$ or $-\frac{5}{2}$
15 $x = 0$ or 1
16 $x = 0$ or $\frac{2}{3}$

2 Unfactorised quadratic equations (pp. 50–51)

1 $x = -2$ or -3
2 $x = -2$ or 5
3 $x = 4$ or 6
4 $x = 0$ or 9
5 $x = 0$ or -3
6 $x = -3$ or 7
7 $x = -7$
8 $x = 4$ or -7
9 $x = 3$
10 $x = 8$ or -3
11 $x = -5$
12 $x = 2$ or -2
13 $x = 9$ or -9
14 $x = 6$
15 $x = 15$ or -4
16 $x = 3$ or 8
17 $x = \frac{2}{3}$ or -4
18 $x = -\frac{2}{5}$ or 7
19 $x = -2$ or $\frac{1}{2}$
20 $x = -\frac{2}{3}$ or $\frac{1}{5}$

3 Quadratic equations where rearrangement is needed (p. 52)

1 $x = 10$ or -9
2 $x = 9$ or 3
3 $x = 2$ or -7
4 $x = 3$ or 4
5 $x = 12$ or -4
6 $x = 4$ or -9
7 $x = 15$ or -4
8 $x = 7$ or 5

Forming and solving quadratic equations (pp. 53–61)

1 Problems about numbers and ages (pp. 53–55)

Try for yourself
$x^2 + 15x - 100 = 0$
$x = -20$ or 5, so she thought of 5.

1 $(x + 4)^2 - 5 = 31$
He could have thought of 2 or -10.
2 $4x^2 = 100$
She could be thinking of 5 or -5.
(Must have both answers.)
3 $2x(2x + 2) = 168$
$x = 6$ or -7, so he was thinking of 12.
4 $x^2 - 6x = 16$
$x = 8$ or -2, so he started with 8.
5 $x(x + 5) = 84$
$x = 7$ or -12, so he thought of 7.
6 $x^2 = 5x + 36$
$x = 9$ or -4, so she started with 9.
7 $x(x + 8) = 3(x + (x + 8))$
$x = 4$ or -6. She thought of 4 and 12.
8 $(a + 4)(a + 7) = 238$
$a = 10$ or -21, so Amber is 10.
9 $(p - 10)(p + 22 - 10) = 135$
$(p - 10)(p + 12) = 135$
$p = 15$ or -17, so Peter is currently 15.
10 $(J - 9)(J + 15) = 145$
$J = 14$ or -20, so they are 14 and 38.
11 $L(L + 2) + 52 = (L + 2)(L + 4)$
$L = 11$, so Liam is 11, and the other two brothers are 13 and 15.
12 $(W + 3)(W + 8) = 150$
$W = 7$ or -18, so Wiremu started with 7 lollies.

2 Problems about lengths, areas and volumes (pp. 56–58)

Try for yourself
$20x^2 - 6x^2 = 350$
$x = \pm 5$; x cannot be -5, so the big rectangle is 25 cm by 20 cm.

1 $x(x + 5) = 204$
$x = 12$ or -17, so the rectangle is 12 cm by 17 cm.
2 $2y \times y \times 5 = 360$
$y = \pm 6$, so the box is 12 cm by 6 cm by 5 cm.
3 $30y^2 + 60y = 1050$
$y = 5$ or -7, so the box is 15 cm by 7 cm by 10 cm.
4 $\frac{1}{2} \times 2z \times (z + 4) = 45$
$z = 5$, so the base is 10 cm long.
5 $x^2 + 12x - 45 = 0$
$x = 3$ or -15, so the width of each arm is 3 m.
6 $x^2 + 28x - 128 = 0$
$x = 4$ or -32, so the width of each arm is 4 m.
7 Area $= \frac{h + (h + 8)}{2} \times h = 45$
$h = 5$ or -9, so the height is 5 cm.
8 $(x + 7)^2 + x^2 = (x + 9)^2$
$x = -4$ or 8, so the sides are 8 cm, 15 cm and 17 cm.
9 $(x + 3)(x + 6) - x^2 = 54$
$x = 4$ cm, so the rectangle was 10 cm by 7 cm.
10 $2x^2 - 16x - 210 = 0$
$x = 15$ or -7, so the living room is 15 m by 15 m.

ISBN: 9780170438568

3 Challenges (pp. 59–61)

1 $a = 11$ and $x = 2$ or 9
2 $a = 4$ and $x = -2$ or 6
3 $a = 1$ and $x = -3$ or 4
4 $a = 7$ and $x = -1$ or -6
5 $a = 8$ and $x = 4$ **6** $n < 3$ and $n > 5$
7 $-3 < n < 8$ **8** $n < -4$ and $n > 3$
9 $2 < n < 10$ **10** $4n - 1$
11 $n + 1$ **12** $6n$

Exponential equations (pp. 62–63)

Basic exponential equations (p. 62)

1 $x = 3$ **2** $x = 4$ **3** $x = 3$
4 $x = 4$ **5** $x = 1$ **6** $x = 0$
7 $x = 6$ **8** $x = 4$

Using equation rules in exponential equations (p. 63)

1 $x = 4$ **2** $x = 2$ **3** $x = 2$
4 $x = 3$ **5** $x = 4$ **6** $x = 0$
7 $x = 2$ **8** $x = 6$ **9** $x = 9$
10 $x = 2$ **11** $x = 4$ **12** $x = 1$

Simultaneous equations (pp. 64–70)

1 Substitution (pp. 64–65)

1 $x = 6$ $y = 5$ **2** $x = 5$ $y = 11$
3 $x = 5$ $y = 2$ **4** $x = 6$ $y = 16$
5 $x = 20$ $y = 55$ **6** $x = -3$ $y = 11$
7 $x = 3$ $y = -2$ **8** $x = -9$ $y = -3$
9 $x = 13$ $y = -3$ **10** $x = 4$ $y = 26$

2 Elimination (pp. 66–67)

1 $x = 4$ $y = 3$ **2** $x = 2$ $y = 5$
3 $x = 11$ $y = -5$ **4** $x = 2$ $y = 7$
5 $x = -1$ $y = 5$ **6** $x = -2$ $y = 5$
7 $x = 3$ $y = 4$ **8** $x = 7$ $y = 1$
9 $x = -3$ $y = 6$ **10** $x = -2$ $y = 4$
11 $x = 2$ $y = -3$ **12** $x = -11$ $y = -2$

Forming and solving simultaneous equations (pp. 68–70)

1 $t = 3s - 16$
$2s + 3t = 150$
Shorts cost $18 and track pants cost $38.

2 $i = 5j$
$4i + 5j = 17.50$
Juicies cost $0.70 and ice creams cost $3.50.

3 $b = 2g$
$b + g = 27$
There are 18 boys and 9 girls.

4 $m = 2a + 5$
$2m + 4 = 6a$
She holds 19 Merit and 7 Achieved grades.

5 $b = 4s$
$s + 42 = 2b$
Bert has 24 and Sadie has 6 marbles.

6 $5l + 2c = 3.30$
$3l + 4c = 4.50$
Lollipops cost $0.30 and chocolate bars cost $0.90.

7 $p = 3t$
$7p + 12t = 231$
Pohutukawa cost $21 and tree ferns cost $7.

8 $6p + 5a = 5.70$
$4p + 7a = 4.90$
Peaches cost $0.70 and apples cost $0.30.

9 $3p = 2b$
$5b + 4p = 27.60$
Pies cost $2.40 and burgers cost $3.60.

10 $h = 2t + 10$
$5h + 11t = 449$
Hoodies cost $48 and T-shirts cost $19.

11 $7c + 4.5s = 127.5$
$c + s = 25$
They sold 6 boxes of chocolate bars and 19 boxes of soap.

Rearrangement of expressions (pp. 71–75)

1 Where the subject appears once (pp. 71–72)

1 $r = \sqrt{\frac{A}{\pi}}$ **2** $P = \frac{100I}{RT}$
3 $x = \frac{y - c}{m}$ **4** $t = \frac{d}{v}$
5 $h = \frac{2A}{a + b}$ **6** $x = \sqrt{\frac{y - b}{a}}$
7 $h = \frac{2A}{B + b}$ **8** $B = \frac{2A}{h} - b$
9 $h = \frac{3V}{r^2}$ **10** $r = \sqrt{\frac{3V}{h}}$
11 $h = \frac{A - 2\pi r^2}{2\pi r}$ **12** $C = \frac{5(F - 32)}{9}$
13 $a = b$ **14** $a = \sqrt{b}$
15 $a = b^2$ **16** $a = b^2$

2 Where the subject appears twice (p. 73)

1 $a = \frac{6}{2 + b}$ **2** $a = \frac{3b}{5 + b}$
3 $a = \frac{2}{2 - b}$ **4** $a = \frac{7b}{6 + 5b}$
5 $a = \frac{4}{b + 1}$ **6** $a = \frac{3b}{2b + 1}$
7 $a = \frac{bc}{b + c}$ **8** $a = \frac{6b}{b + 3}$

3 Problems involving the rearrangement of expressions (pp. 74–75)

1 $H = 3h$ **2** $B = \sqrt{9b^2 + 288}$ **3** $L = \sqrt{4l^2 - 108}$
4 $h = \frac{4}{3}r$
The height is four thirds the length of the radius.

ISBN: 9780170438568

5 $h = r$
The height and the radius are the same length.
6 $h = \sqrt{3}r$
The height is the radius multiplied by $\sqrt{3}$.
7 $x = \frac{2\pi r}{3}$

Algebraic proof (pp. 76–77)

1 Let three consecutive odd numbers be x, $(x + 1)$ and $(x + 2)$.
$x + (x + 1) + (x + 2) = 3x + 3$
$= 3(x + 1)$
$3(x + 1)$ is always a multiple of 3.
2 Let the two even numbers be $2x$ and $2y$.
$(2x) \times (2y) = 4xy$
$4xy$ is always a multiple of 4.
3 Let the two odd numbers be $(2x + 1)$ and $(2y + 1)$.
$(2x + 1) \times (2y + 1) = 4xy + 2x + 2y + 1$
$= 2(2xy + x + y) + 1$
Since $2(2xy + x + y)$ is a multiple of 2, it must be even.
$\therefore$ $2(2xy + x + y) + 1$ must be odd.
4 Let the smaller number be x and the larger number be $(x + 1)$.
$(x(x + 1)) + (x + 1) = x^2 + x + x + 1$
$= x^2 + 2x + 1$
$= (x + 1)^2$
$(x + 1)^2$ is a square number.
5 Let n be even, and $n = 2x$:
$n^2 + n + 2 = (2x)^2 + (2x) + 2$
$= 4x^2 + 2x + 2$
$= 2(2x^2 + x + 1)$
$2(2x^2 + x + 1)$ is a multiple of 2, so it must be even.
Let n be odd, and $n = 2x + 1$:
$n^2 + n + 2 = (2x + 1)^2 + (2x + 1) + 2$
$= (4x^2 + 4x + 1) + (2x + 1) + 2$
$= 4x^2 + 6x + 4$
$= 2(2x^2 + 3x + 2)$
$2(2x^2 + 3x + 2)$ is a multiple of 2, so it must be even.
$\therefore$ $n^2 + n + 2$ must always be even.
6 $a + b = 9$
$\therefore$ $b = 9 - a$
The two-digit number $= 10a + b$
$= 10a + (9 - a)$
$= 10a + 9 - a$
$= 9a + 9$
$= 9(a + 1)$
$9(a - 1)$ is always a multiple of 9.
7 Let the number be written 'a5'.
It has the value $(a \times 10 + 5) = 10a + 5$
$(\text{'a5'})^2 = (10a + 5)^2$
$= 100a^2 + 100a + 25$
Both $100a^2$ and $100a$ must both be multiples of 100, so their sum will also be a multiple of 100 and end with two zeros.
$\therefore$ any square of a whole number under 100 and ending in 5 must end in 25.

General pattern and formula problems (pp. 78–81)

1 a 420 dots
b If $D = 110$, $n = 10$ or -11. Since the pattern number cannot be negative, it must be pattern number 10.
2 a \$120
b 3 coats
c $C = \frac{11T}{LWP}$
3 a 12 b 21
c $c = 3M - a - b$
4 a $T = 300 + 0.1(x - 2000)$
b \$460 c \$5800
d $x = 2000 + \frac{T - 300}{0.1}$
5 a 211 matchsticks
b If $M = 46$, $n = 9$ or -10. Since the pattern number cannot be negative, it must be pattern number 9.
c $T = 4M$
6 $y = 21n - 20$: 1, **22, 43**, 64
$y = n^3$: 1, **8, 27**, 64
$y = 4^{n-1}$: 1, **4, 16**, 64
7 a $S = 76\text{ cm}^2$
b Width = 5 cm
c $l = \frac{s - 2wh}{2w + 2h}$
8 a If $M = 90$, $n = -11$ or 8. Since the pattern number cannot be negative, it must be pattern number 8.
b $T = 5M$

Algebra with rectangles (pp. 82–83)

1 a $P = 6x + 10$
b $x = 3 \rightarrow$ rectangle is 9 cm by 5 cm
c $A = 2x^2 + 7x + 6$
2 a $P = 10x + 2$
b $x = 5 \rightarrow$ rectangle is 17 cm by 9 cm
c $A = 6x^2 + x - 2$
3 a $\frac{x^2 + 2x - 3}{x - 1} = x + 3$
b $x = -11$ or 9. Cannot have a negative dimension $\rightarrow x = 9$
Dimensions: 12 cm by 8 cm
4 a $\frac{x^2 + x - 20}{x - 4} = x + 5$
b $x = -8$ or 7. Cannot have a negative dimension $\rightarrow x = 7$
Dimensions: 3 cm by 12 cm

5 **a** $p = 3x + 1$ $q = x - 1$
b $P = 12x - 2$ **c** $x = 3$
d $A = 7x^2 - x - 2$

6 **a** $p = x + 1$ $q = 3x - 8$
b $P = 12x - 14$ **c** $x = 5$
d $A = 5x^2 - 9x + 8$

Practice questions (pp. 84–89)

Practice question 1 (pp. 84–85)

a **i** $(2x + 5)$ and $(x - 4)$
ii If the product is 0, then $x = -\frac{5}{2}$ or $x = 4$.
For the product to be positive,
then $x < -\frac{5}{2}$ or $x > 4$.

b **i** $r = \sqrt{\frac{V}{\pi h}}$
ii $r = \sqrt{\frac{54\pi}{6\pi}} \Rightarrow r = 3$

c **i** 2, $(2x + 1)$ and $(x - 3)$
Or $(2 + 0x)$, $(2x + 1)$ and $(x - 3)$
ii 18 units^3

d Let the number be 'xy'.
The value of 'xy' is $10x + y$.
$\therefore (10x + y) - (x + y) = 9x$.
$9x$ is a product of 9 and x, so this always works.

Practice question 2 (pp. 86–87)

a Length $= x^2 - x - 2$
b Length $= 4x^3y$
c $x = \pm\sqrt{2}$
d She could be thinking or 5 or –12.
e $A = 3B$
f $3a = 5c$
$a + 1.70 = 2c$
Adults' tickets cost \$8.50 and children's tickets cost \$5.10
g $p = -4$

Either demonstrate that solution produces a contradiction such as $0 = 2$
or demonstrate an understanding that $p = -4$ would result in two parallel but distinct lines that can never meet, so there are no solutions.

Practice question 3 (pp. 88–89)

a **i** 55
ii $n = 9$
iii $\frac{1}{2}n(n + 1) = 50$
$n(n + 1) = 100$
$\therefore$ two consecutive whole numbers must be able to be multiplied $\Rightarrow$100.
The nearest numbers are $9 \times 10 = 90$, and $10 \times 11 = 110$.
$\therefore$ there are no pairs of consecutive whole numbers that satisfy this equation, so she could not make the pattern with whole groups and no buttons left over.

b
$$\frac{2(3p + 2)}{6} - \frac{1}{p} = \frac{2(3p + 2)}{6} \times \frac{p}{p} - \frac{1}{p} \times \frac{6}{6}$$
$$= \frac{6p^2 + 4p}{6p} - \frac{6}{6p}$$
$$= \frac{6p^2 + 4p - 6}{6p}$$
$$= \frac{3p^2 + 2p - 3}{3p}$$

c $2^6 = 64$ and $2^7 = 128$.
$\therefore m$ must be 7.
So $2m - 3 = 7 \Rightarrow m > 5$

d $g = \frac{P + 11}{14}$ or $\frac{P - 45}{14} + 4$

e
$$2 - \frac{5}{x} = 7y$$
$$2x - 5 = 7xy$$
$$2x - 7xy = 5$$
$$x(2 - 7y) = 5$$
$$x = \frac{5}{2 - 7y}$$

ISBN: 9780170438568